geo planepuzzles

Exploring Properties of 2D Shapes

ALLAN TURTON
CALVIN IRONS

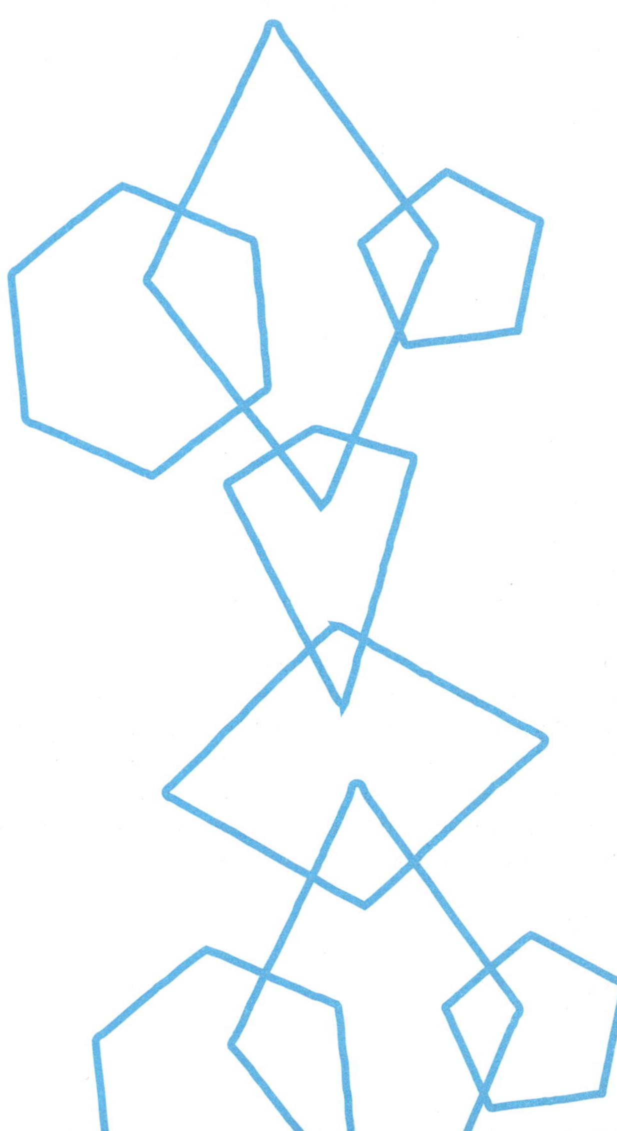

GEO Plane Puzzles

Copyright 2006 ORIGO Education
Authors: Allan Turton and Calvin Irons

Turton, Allan.
GEO plane puzzles : exploring properties of 2D shapes.

ISBN 1 876842 87 3.
1. Geometry - Study and teaching (Primary). I. Title.

372.7044

For more information, contact
North America
Tel. 1-888-ORIGO-01 or 1-888-674-4601
Fax 1-888-674-4604
sales@origomath.com
www.origomath.com

Australasia
For more information,
email info@origo.com.au
or visit www.origo.com.au for other contact details.

ISBN: 1 876842 87 3

10 9 8 7 6 5 4 3 2 1

contents

Geometry

Geometry is the exploration of space, size, and position. Our knowledge or intuitive understanding of geometry plays a major part in our everyday and professional lives. For example, a simple task of reading a map requires an understanding of direction and position. A more demanding task of repairing an engine draws on an intuition about shapes and the effect of moving them. The mechanic is required to visualize the shapes and how they fit together. Likewise scientists, engineers, and carpenters are all required to know and use certain geometric ideas that are specific to their profession. For this reason alone, it is easy to see why geometry has become an essential part of the school mathematics curriculum. However, there is an even more important reason for placing greater emphasis on the teaching of geometry. Geometry involves the manipulation of mental pictures, which is often called visual thinking. Problem solving in all strands of mathematics depends on forming mental pictures of the situation in which the problem is embedded and then "finding" a picture of the mathematical idea that matches. The ability to mentally form, rearrange, and match pictures is crucial to all aspects of mathematics, particularly problem solving.

Did You Know?
The term "geometry" simply means "earth measure".

The *GEO* Series

These *GEO* books are designed to help teachers make geometry an essential component of their mathematics curriculum. The eight books provide hundreds of exciting and innovative activities for teaching geometry to students aged 6 to 12 years and beyond. The topics include two-dimensional shapes; three-dimensional shapes; lines and angles; symmetry; tessellation; topology; folding two-dimensional shapes; and location, direction, and movement.

Each book contains easy-to-read instructions and fully reproducible blackline masters. Each activity has two types of icon: one indicates the appropriate age group for the activity and the other indicates whether the activity is designed as a whole class, small group, pairs, or individual activity. Definitions of terms are provided as friendly reminders and interesting facts appear in the margin. A glossary is also provided at the back of each book.

Those teachers who use all eight books will develop a greater understanding of geometry and new insights into its use. Selecting and using activities from the series will offer students a broad view of geometry and one that is inviting, engaging, and thought-provoking.

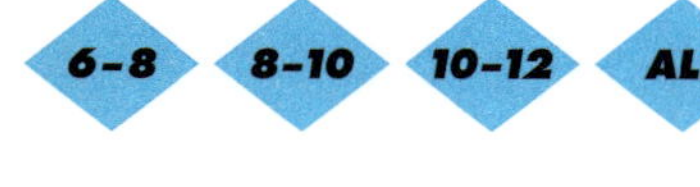

▲ *These icons show the appropriate age group for the activity.*

▲ *These icons show whether the activity is designed for an individual, a pair, a small group or a whole class.*

About *GEO Plane Puzzles*

This book contains a variety of exciting ideas to explore two-dimensional shapes. It provides activities based on describing, sorting, representing, and altering a variety of shapes. Particular focus is given to curves, triangles, quadrilaterals, and regular pentagons.

Suggested Materials

A range of materials can be used to support the activities in this book. Some activities will use everyday craft items such as blank paper, drinking straws, pipe cleaners, toothpicks, playdough, string, and similar objects. Others require particular classroom materials including pattern blocks, geostrips, and drawing compasses. Provided at the back of *Plane Puzzles* are blackline masters for specialty items such as playing cards, tangram puzzles, and speciality grid paper.

Assessment

Activities in the *GEO* series have been designed to help students demonstrate key outcomes and standards in geometry. An assessment chart has been designed to identify those activities that can help you assess your students' understanding and achievement. The chart is available from the following websites:

North America — www.origomath.com

Australasia — www.origo.com.au

Terminology

Most terms used to describe geometric shapes are the same in all English-speaking countries. However, some terms common in one place are not in another. One example involves naming types of rectangles. Some students find it confusing that squares are described as types of rectangles. The term "non-square rectangle" is then sometimes used to describe those rectangles that are not squares. A more concise way of describing such shapes is to call them "oblongs". Using this term, an analogy can be drawn between squares and oblongs being types of rectangles in the same way that a sedan and a limousine are both types of cars. For these reasons, "oblong" has been used in this book.

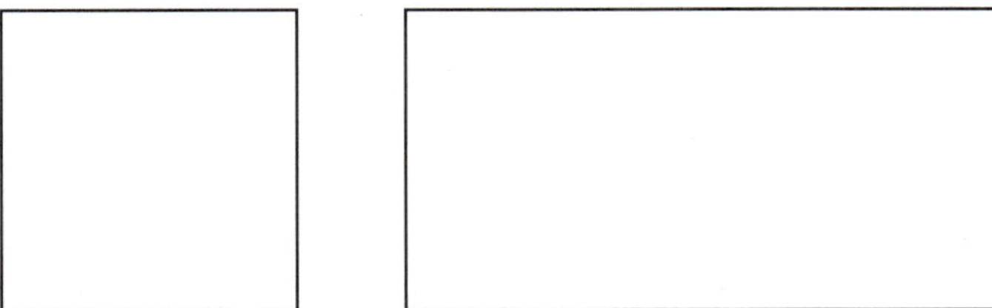

▲ *Both of these shapes are rectangles, but in this book the shape on the right is called an oblong.*

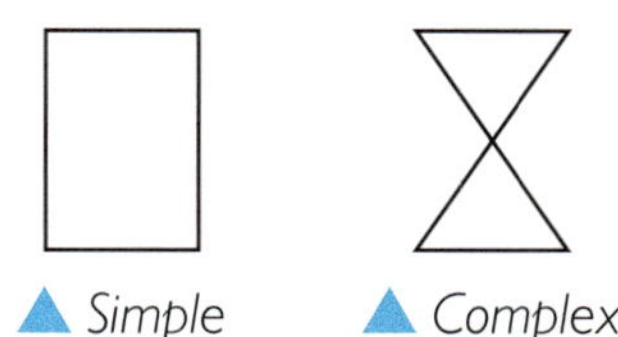

A polygon is any simple two-dimensional closed shape formed by three or more straight line segments.

A quadrilateral is any polygon with four sides.

A similar example can be found when talking about parallelograms where the pairs of parallel sides are of different lengths and there are no 90° angles. Often such shapes are just called parallelograms but a more precise term to use would be "rhomboid". However, this word is not well known and can often be confused with "rhombus", so "parallelogram" is the term used in this book.

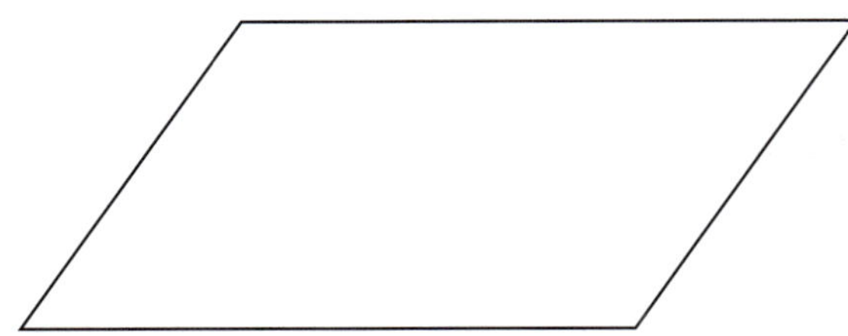

▲ *In this book, the shape above is called a parallelogram.*

A third example is the quadrilateral with only one pair of parallel sides. In North America, it is called a trapezoid while in Australia, Europe, and elsewhere this same shape is called a trapezium. In this book, the term "trapezoid" is used to refer to the shape.

▲ *In* Plane Puzzles, *these shapes are called trapezoids.*

Finally, a line is traditionally thought of as extending infinitely into space, while a line segment is just a part of that line. For ease of reading, this book generally uses "line" to describe both lines and line segments. Teachers may like to model the correct terminology for their students, but it is not overly important for students to remember it.

Two-Dimensional Shapes

This book is all about investigating "plane shapes". A plane can be thought of as a flat surface without thickness that extends in all directions infinitely. A plane shape is one that lies on that plane and is completely flat. Unfortunately, we can never hold or construct a true plane shape because even our most true-to-life models of a plane, such as a sheet of paper, have thickness.

Another way to describe plane shapes is to say that they have two dimensions, or that they are "two-dimensional". There are several ways to define dimensions in order to discuss plane shapes:

1. One use of "dimensions" is to describe the shape as having certain "properties". In this sense, a cube has three dimensions (length, width, and height), a square has two (length and width), and a line segment has only one (length). The difficulty with this definition is that one- and two-dimensional shapes can be described in words but not accurately drawn with pictures or made with materials. Using "dimensions" as a type of property to describe a square would mean that a "two-dimensional" square constructed from drinking straws would actually be "three-dimensional" because the straws have length, width, and height. So using "dimensions" in this way could be a confusing way of thinking about shapes.

Name of polygon	No. of sides
triangle	3
quadrilateral	4
pentagon	5
hexagon	6
heptagon	7
octagon	8
nonagon	9
decagon	10
hendecagon	11
dodecahedron	12

▲ *Blackline Master 6 on page 64 reveals the Greek and Latin origins of the names for many of the polygons discussed in* Plane Puzzles.

2. A different way of thinking about the word "dimensions" is to use it to describe the lineal measurements we usually use to describe a shape. In this example, "dimensions" are not properties of a shape, but measurements in a particular direction. Using this definition of "dimensions", a prism is usually measured using three dimensions (length, width, and height). A triangle, square, and even a circle are usually measured using two dimensions (length and width or similar), while a line segment is usually measured using only one dimension (length).

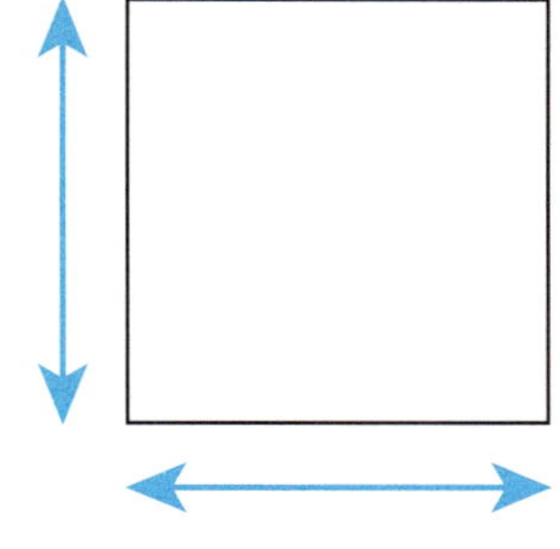

▲ *A square is usually measured using two dimensions that happen to have the same value.*

3. The most precise approach in using "dimensions" as measurement is to consider how many dimensions, or particular measurements, would be needed to identify the position of a point on or inside any geometric figure. To locate a point anywhere along a line segment, all that is required is to know the distance from one end of the line segment to where the point is. If we think of a triangle lying on a co-ordinate grid, then two dimensions can be used to describe a point anywhere on or inside the triangle. Similarly, any point on or inside a cube can be found using three measurements from, for example, one of the corners on the base of the cube.

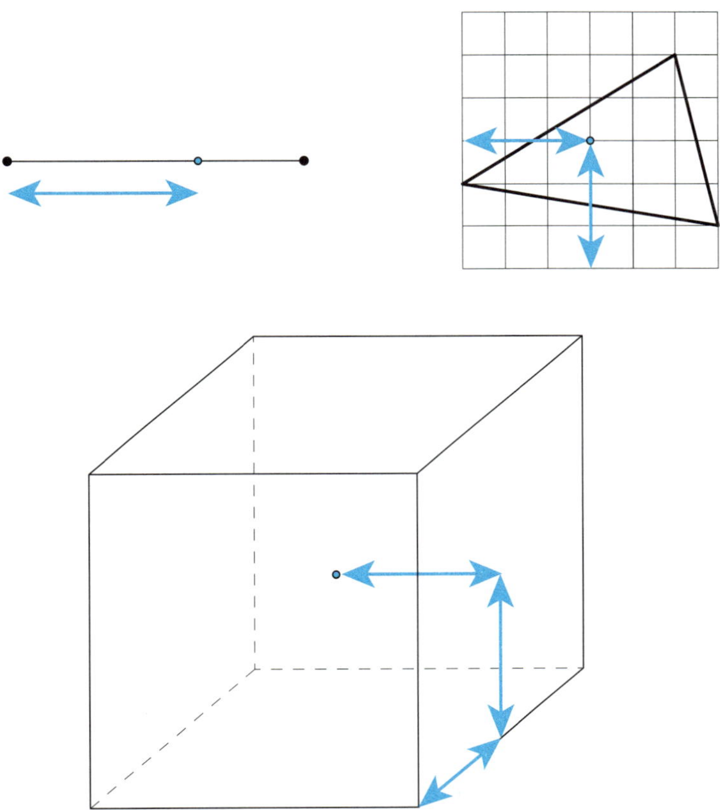

▲ *Different numbers of dimensions are needed to describe the position of a point on these objects.*

In summary, any concrete model of a two-dimensional or plane shape will have thickness. Students may better understand the concepts involved with this area of geometry if they consider dimensions as measurements and think of a two-dimensional shape as being one for which two dimensions are required to locate a point on or inside the shape.

Classifying two-dimensional shapes

For your reference, the classification chart below shows how certain two-dimensional shapes are related to one another. Explanations of each term can be found in the glossary at the back of this book.

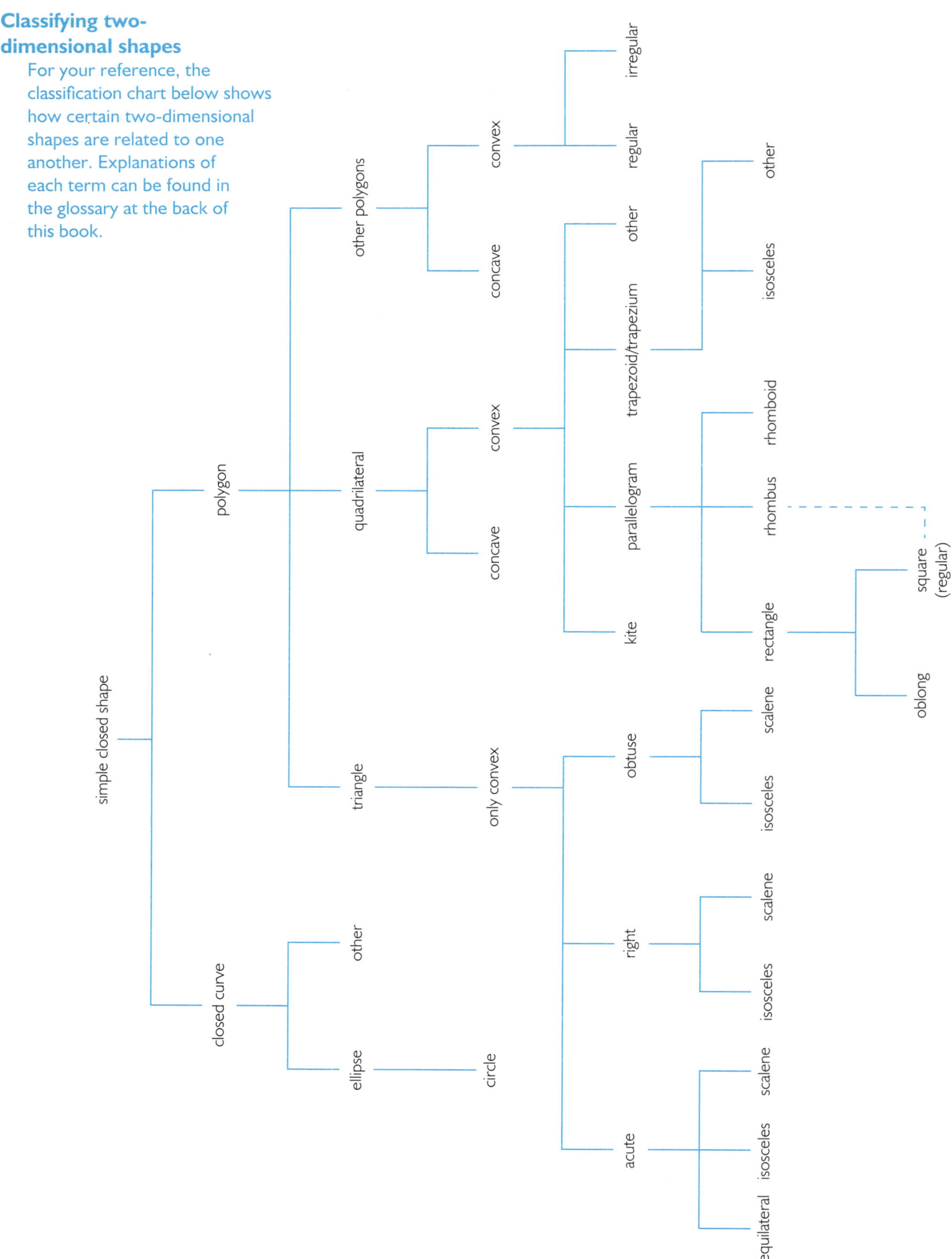

▲ *A classification chart for simple, closed two-dimensional shapes.*

all shapes — describing

From wire coat hangers to the faces of pyramids, representations of two-dimensional shapes can be seen everywhere around us. The following activities are designed to help the students recognize a range of two-dimensional shapes and identify their defining properties. The playing cards listed in Activity 3 are used extensively in this unit and others, so it may be worthwhile laminating them.

1. Finding Shapes

Preparation
No preparation is required.

Activity

1. Take the students for a walk around the school and ask them to point out any two-dimensional shapes they recognize. For example, they can look for oblongs in brick faces or circles on basketball courts. Other examples of shapes found in built and natural environments include road signs, honeycombs, sporting fields and courts, buildings, fabric patterns, money, and the faces of polyhedra (three-dimensional shapes).

2. Provide the students with magazines and have them cut out pictures of familiar two-dimensional shapes. The students can use the plain paper to make a "shape book" as a record of the shapes they have seen and where they saw them.

3. To extend this activity, ask the students to find and record non-examples of shapes and explain why they think each shape is a non-example.

■ Materials
- Magazines
- Scissors — 1 pair for each student
- Glue — for each student
- Plain paper — several sheets for each student

2. Word Wall

Preparation
No preparation is required.

Activity

1. As the students learn the names of particular shapes, they can write each name and draw a picture of the shape on a card. These cards can then be arranged on a blank wall.

2. Pictures of real-life examples of the shapes can also be added to the wall, and arranged around the corresponding words.

■ Materials
- 7.5 cm × 15 cm blank cards — several for each student

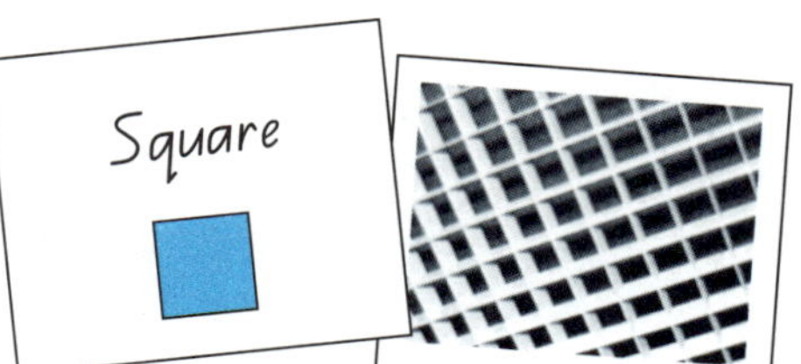

▲ The students can use the word wall as a reference tool.

■ Materials

- Blackline Masters 1 to 5 (pages 59–63)
- Extra tagboard or light card cut into cards of the same number and size as the blackline master cards

Did You Know?

In North America, the term "trapezoid" is used to describe a quadrilateral that has only two parallel sides. In Australia, Europe, and elsewhere, the same shape is called a "trapezium".

An oblong is a rectangle with adjacent sides of different lengths. It is also known as a "non-square rectangle".

3. Card Games

Preparation

1. Copy Blackline Masters 1 to 5 onto tagboard or light card and cut out the cards.

2. On the extra cards, write a description for each of the blackline master cards. The descriptions can be guided by the attributes of the shapes you want the students to focus on. (Suggestions by age group are given below.) This will help the students make verbal and visual connections between the shapes and their names.

Activity

1. Card games such as "Concentration" (also known as "Memory") and "Go Fish" can be played by the students with the cards from Blackline Masters 1 to 5.

2. The students can next play the games by using only shapes, using only words (from the pair cards you have created), or using half and half.

 Suggestions include:
- *Use the names and shapes of triangles, oblongs, squares, and circles.*
- *Use shapes described as having "5 sides" and "6 sides", or "5 corners" and "6 corners".*

 In addition to the above:
- *Use the names and shapes of quadrilaterals, parallelograms, rhombuses, trapezoids, and kites.*
- *Use the names and shapes of equilateral, isosceles, and scalene triangles.*
- *Use the names and shapes of pentagons, and hexagons.*
- *Use the terms "parallel", "regular", and "irregular".*

 In addition to the above:
- *Use the terms "convex" and "concave".*

4. Can You Do It?

Preparation

No preparation is required.

Activity

1. Challenge the students with a range of possible and impossible tasks. Direct the students to use the geoboards, square-corner testers, or drawings to display their answers. Choose tasks appropriate for the students' ages. Ask younger students questions such as, *Can you make a triangle with only three corners?* (Yes.) *Can you make a square with five sides?* (No.) Ask older students questions such as, *Can you make a rectangle with an angle of 80°?* (No.)

2. As the students consider the questions, and experiment as needed, ask them to explain their answers. Their explanations will reveal how well they understand the concepts surrounding a particular shape.

■ Materials

- Geoboard and rubber bands or similar (e.g. drinking straws and pipe cleaners) — 1 set for each student
- Square-corner tester — 1 for each student (see below)

▲ *Have the students make square-corner testers to examine shapes.*

5. What Is It?

Preparation

No preparation is required.

Activity

1. Suggest a property of a shape to the students and direct them to construct a shape that has that property. For example, if you say, *This shape is a closed shape,* the students can construct any shape, providing it is closed. Have the students construct three or four examples of the shape you describe.

2. Give the students additional clues to limit their choice of shapes. For example, say, *This shape also has straight sides,* so that the shape must be a polygon. Then say, *The shape has only five angles,* so that the shape must be a polygon with only five angles, that is, a pentagon.

6–8 Use triangles (including equilateral triangles), rectangles (including squares and oblongs), pentagons, and hexagons. The properties can include whether the shape is open or closed, the number of sides, the number of corners, and the number of square corners.

8–10 In addition to the above shapes, use octagons, rhombuses, trapezoids, and kites. The properties can also include whether a shape is regular or irregular, and whether it has acute, obtuse, or right angles.

10–12 In addition to the above shapes, use ellipses, isosceles triangles, and scalene triangles. The properties can also include whether a shape is convex or concave.

■ Materials

- Geoboards and rubber bands or similar (e.g. drinking straws and pipe cleaners) — 1 set for each student
- Square-corner tester — 1 for each student

Did You Know?

The interior angles of a convex two-dimensional shape are all less than 180°.

Concave two-dimensional shapes have at least one interior angle that is greater than 180°.

The shape has straight lines.
(This means it cannot have any curves.)

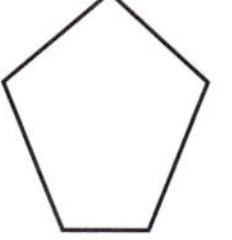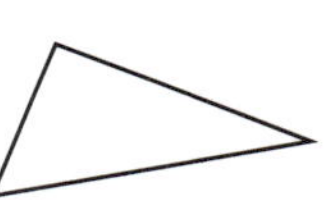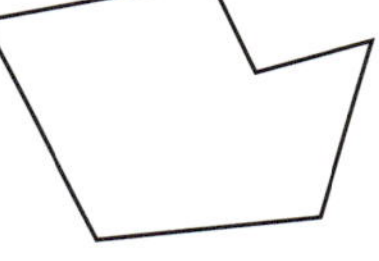

The shape has five sides.
(This means it can only be a pentagon.)

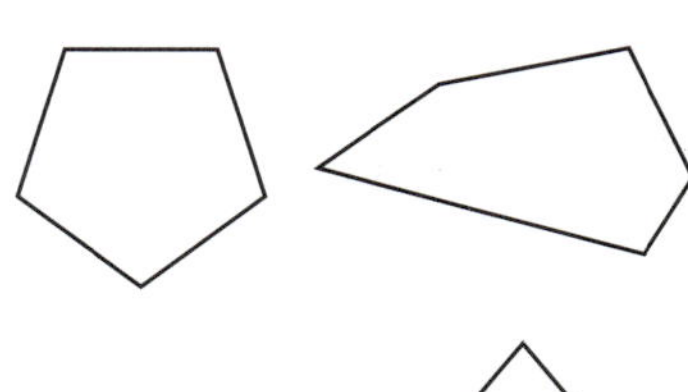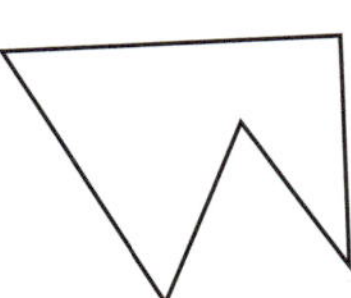

The shape has two right angles.
(Two right angles must be indicated.)

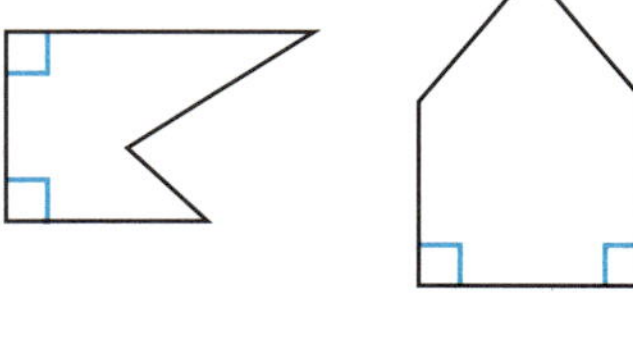

The shape is concave.
(Two right angles and the concave angle must be indicated.)

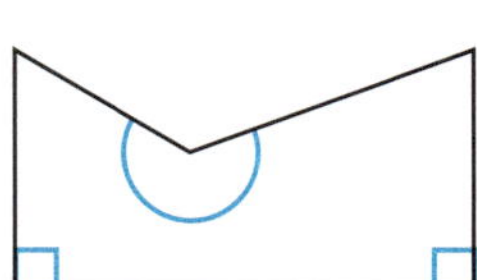

▲ *Given the above sequence of clues, older students may draw shapes like these.*

Materials

- Drinking straws and pipe cleaners — for each student and for demonstration
- Lengths of string — for each student and for demonstration

6. Open and Closed Shapes

Preparation

Make a number of shapes such as these below.

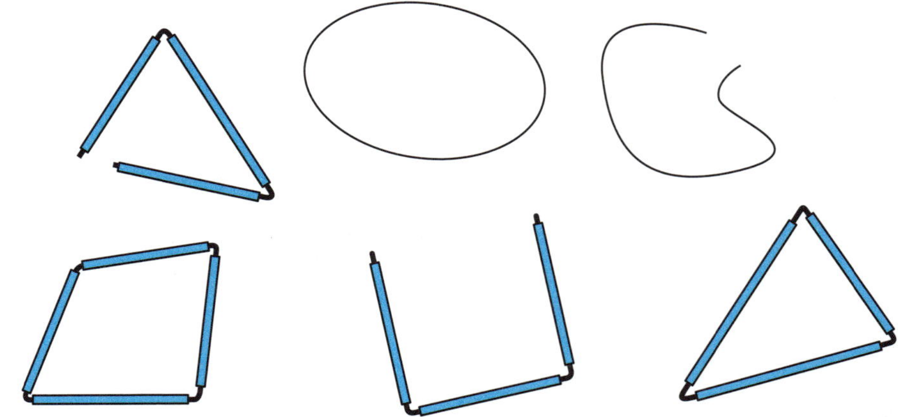

▲ *The students should be able to describe how some shapes are "open", while others are "closed".*

Activity

1. Display the shapes you prepared, so that the students can see them. Choose two of the shapes. Encourage a volunteer to describe any similarities and differences they can see between the shapes.

2. Repeat with another two shapes, then invite volunteers to nominate shapes they think are "similar" or "different", and explain why.

3. Introduce the terms, "closed", to describe shapes where each "end" of each side connects to another, and "opened", to describe shapes where not all the "ends" of each side join others.

4. Have the students make three closed shapes and three open shapes.

Materials

- Blackline Masters 1 to 5 (pages 59–63)

7. Three Facts

Preparation

Copy Blackline Masters 1 to 5 onto tagboard or light card and cut out the cards (or reuse the cards made in Activity 3).

Activity

1. Hold up one of the cards and say, *Tell me three facts about the shape on this card.* Some students may talk about how many corners some shapes have; others may talk about the number of sides; while others may know the shape names. Descriptive language can be informal at this stage. The purpose is to have students start thinking and talking about the properties of the shapes.

2. After this whole class activity, you may like to have the students work in small groups or individually to examine a given card.

8. Finding Square Corners

Preparation

No preparation is required.

Activity

1. A handy and uncomplicated tool that can be used to find right angles in two-dimensional shapes is a "square-corner tester". To make the tester, instruct the students to tear off the corners of the white sheet of paper, and fold the remaining paper in half, and then in half again. Have the students paste the small square of color paper where the center folds are to show the square corner.

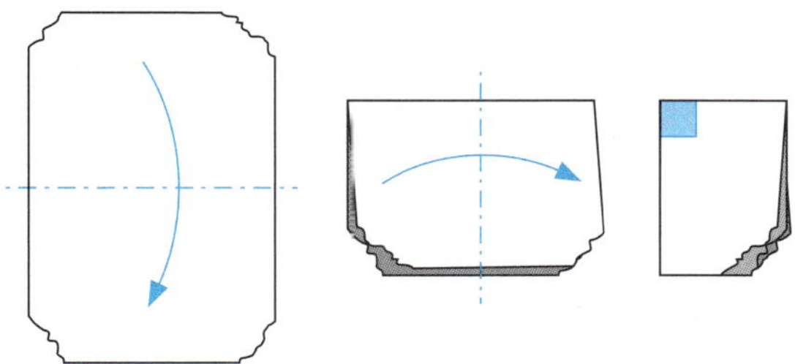

▲ *Have the students use square-corner testers to examine 2D shapes around the room.*

2. To help the students understand why it is called a square-corner tester, complete Activity 1 ("What is a Square?") on page 45.

3. Challenge the students to find shapes or other features in the room that show a square corner. Discuss the students' findings as a class.

9. Regular and Irregular Polygons

Preparation

1. Copy Blackline Masters 1 to 5 onto tagboard or light card and cut out the cards. You need one set for each group of students.

2. Cut the overhead transparencies (OHTs) into eighths and give one piece to each group of students.

Activity

1. Direct small groups of students to sort the cards into two piles: shapes with all sides the same length and those with sides of differing lengths. (For example, a square has four sides of the same length.)

2. Say, *Examine the cards that have shapes with sides the same length. Sort them into another two piles: shapes with all angles the same size and those with angles of differing sizes.* Many students will be able to determine this by estimation, but some may need another way of analyzing, for example, the use of a tool. To help check angle sizes, the students can place a piece of OHT over an angle in one of the shapes and then trace that angle. The traced angle can then be positioned over the remaining angles in the shape to test if they are the same.

Materials

- Sheet of white paper — 1 for each student
- 2.5 cm square of color paper — 1 for each student
- Glue

geo

Dozens of interesting activities can be found in *All About Angles*.

Materials

- Blackline Masters 1 to 5 (pages 59–63)
- Overhead transparency sheets — enough sheets to cut into eighths, with 1 piece for each group of students
- Overhead transparency pens — 1 for each group of students

3. Explain that polygons that have all sides equal in length and all angles equal in size are called "regular polygons". Discuss how some have special names like "square" and "equilateral triangle", but most just have the word "regular" in front of their names (e.g. "regular pentagon"). Tell the students that all other types of polygons can be called "irregular".

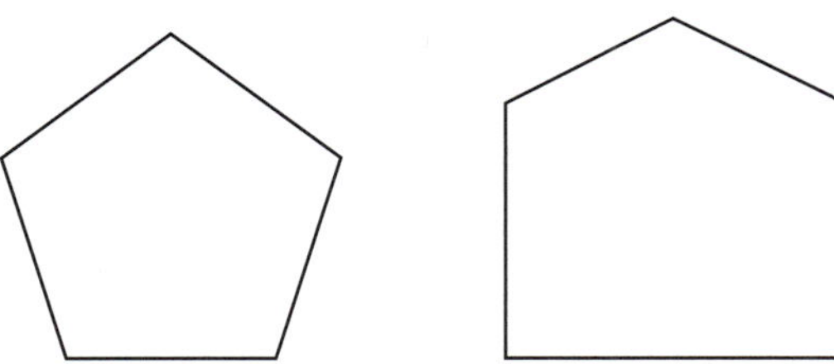

▲ *After measuring, the students should be able to explain why the shape on the left is regular and the one on the right is irregular.*

■ Materials

- **Geostrips and fasteners or similar** (e.g. drinking straws and pipe cleaners) — 1 set for each student

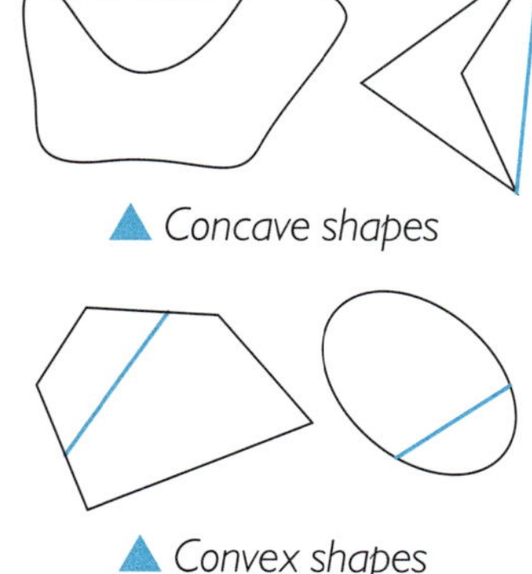

▲ Concave shapes

▲ Convex shapes

10. Concave and Convex

Preparation

Draw the following shapes on the board.

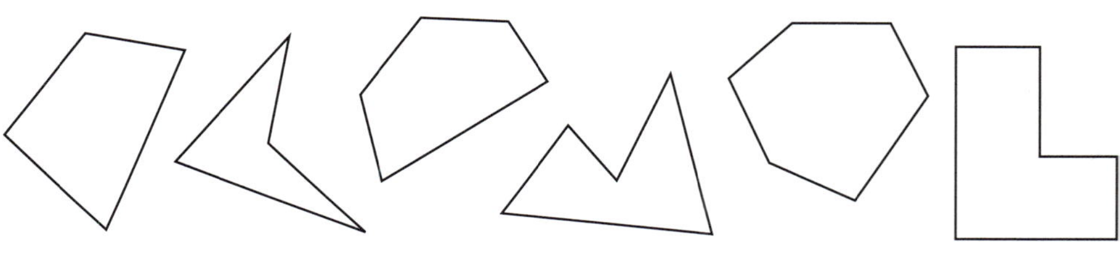

▲ *The students can identify the largest angle in each of these shapes to help them understand the difference between concave and convex shapes.*

Activity

1. Point to the quadrilaterals and ask, *What is the same about these two shapes?* (They both have four straight sides.) *What is different about them?* Answers may vary and include ideas such as the concave shape being "pointy" and "almost a triangle".

2. Repeat with the pair of pentagons and the pair of hexagons. Encourage the students to consider the size of the interior angles of the shapes. Ask, *What is the largest interior angle that you can see in each of the shapes? How could you describe how large they are? Are they more or less than a right angle? Are they more or less than a straight angle?*

3. Explain that *concave* shapes have at least one interior angle that is larger than 180°. Another way of describing this is to say that a *concave* shape looks like it has been "pushed in". *Convex* shapes, on the other hand, do not have any interior angles that are larger than 180°.

5. Direct the students to use the geostrips to construct examples of concave and convex shapes.

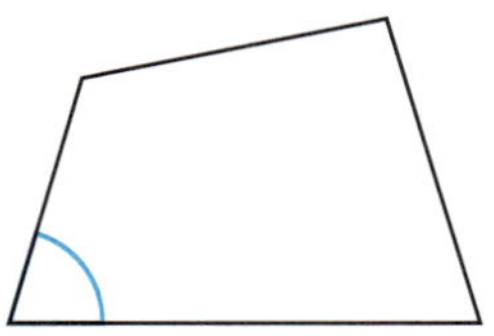

▲ An interior angle lies inside a shape.

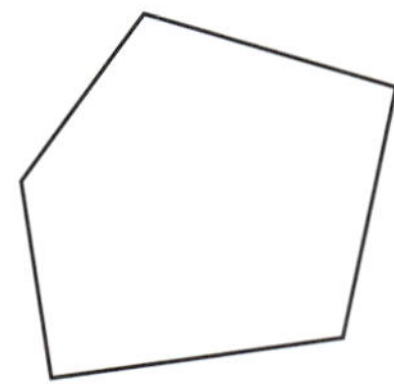

▲ Convex shapes have no interior angles that are larger than 180°.

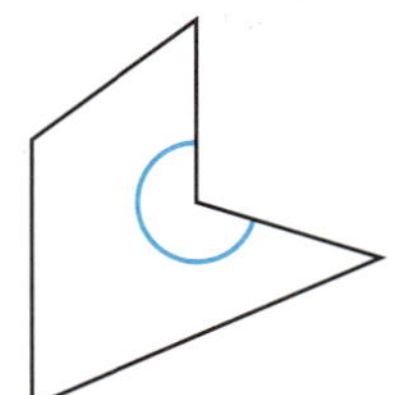

▲ Concave shapes have at least one interior angle that is larger than 180°.

11. Roots

Preparation

Make one copy of Blackline Master 6 for each student.

Activity

1. Discuss the origins of the words used to describe two-dimensional shapes. For example, the Latin root *quadri-,* means "four", and is evident in the related words, "quadrilateral" and "quadrant".

2. Select other words that are used to describe two-dimensional shapes (Blackline Master 6 provides some examples). Write the root word or words, with the meanings, on one of the blank cards and stick the card on the board. As the students use the dictionaries to find other words (including non-mathematical ones) that are related to the root word, they can write them on the blank cards and then arrange them around the original card.

■ Materials

- Blackline Master 6 (page 64)
- Dictionary — 1 for each student
- A teacher's dictionary that shows the origins of words
- Blank cards, approx. 7.5 cm × 12.5 cm
- Blu-Tack

12. Definitions

Preparation

No preparation is required.

Activity

1. Direct the students to each write a definition for one of the shapes they have studied, on a piece of scrap paper. Tell them that their definitions should not include the shape name.

2. Arrange the students into groups. Each student in turn reads his or her definition to the group. The other students then guess what the shape is by constructing a shape that matches the definition.

3. Once the shape is identified (either by the group or the author), the other students in the group then try to make other shapes that meet the definition but are not the intended shape. Encourage discussion as to how the definition can be refined if more than one shape is possible.

4. Repeat the process for all students in the group. The glossary on page 79 describes the key features of many shapes.

■ Materials

- Geostrips and fasteners or geoboards and rubber bands — 1 set for each group of students
- Scrap paper — 1 sheet for each student

all shapes — sorting

Sorting and categorizing are processes undertaken by people of all ages. The activities in this unit are designed to help students consider the similarities and differences between shapes, and how the shapes might be classified. The playing cards listed in the first activity are used extensively in this unit and others, so it may be worthwile laminating them.

■ Materials
* Blackline Masters 1 to 5 (pages 59-63)

1. Odd One Out

Preparation
Copy Blackline Masters 1 to 5 onto tagboard or light card and cut out the cards (or reuse the cards made in previous activities).

Activity

1. Show the trapezoid, rhombus, square, and regular pentagon cards. Ask, *Which one of these shapes is not like the others? Why?* A variety of answers may be offered by the students, depending on their age. Younger students, for example, may say, *The one with five sides is the odd one out because all the others have four sides.* Older students may say, *The trapezoid is the odd one out because all the others have sides that are equal in length,* or, *The square is the odd one out because it's the only one with a right angle.* There may be more than one correct answer (in fact, it is better if there *is* more than one). Ask the students to give reasons that a particular shape is the odd one out.

2. Arrange the students into groups of three and distribute the cards so that each group has about four or five cards. Ask the students to consider which shape is different from the others. After they discuss their ideas as a group, have them write or talk about what they thought. Invite volunteers to share their ideas with the whole class.

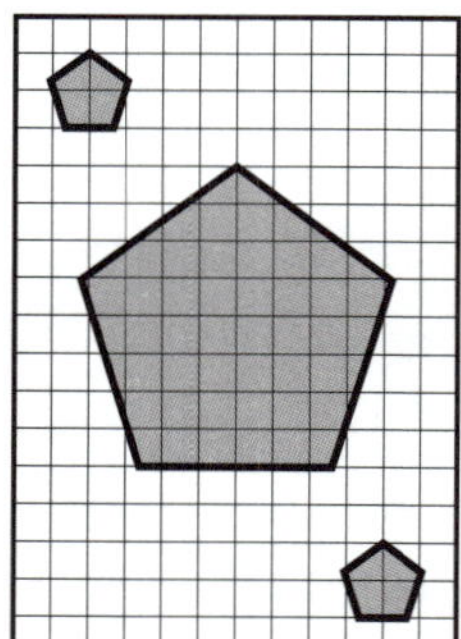 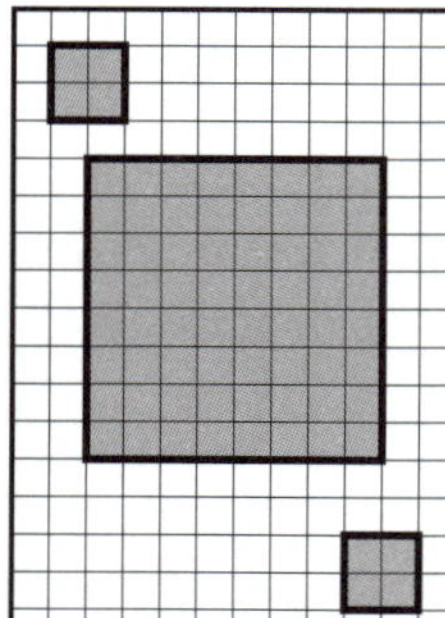 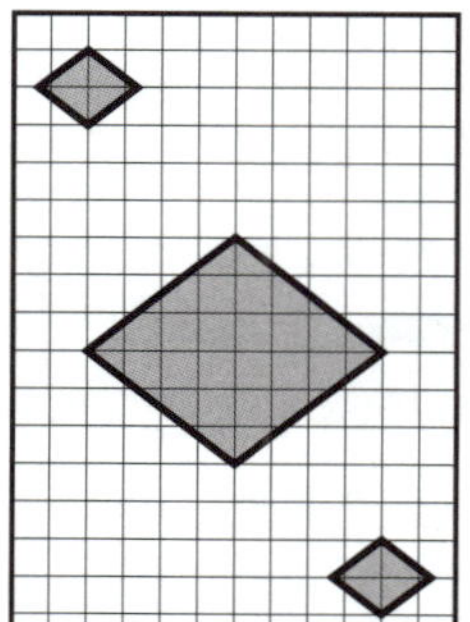 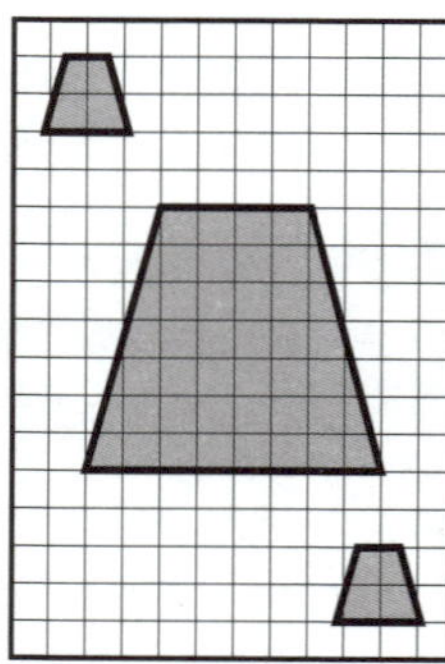

▲ *Ask the students to offer reasons why one shape is not like the other shapes.*

Did You Know?
A regular polygon has equal angles and equal sides.

Did You Know?
In North America, the term "trapezoid" is used to describe a quadrilateral that has only two parallel sides. In Australia, Europe, and elsewhere, the same shape is called a "trapezium".

2. Sorting Shapes

Preparation

Copy Blackline Masters 1 to 5 onto tagboard or light card and cut out the cards (or reuse cards made in Activity 1).

Activity

1. This activity is helpful in determining how students examine shapes. Choose a selection of the cards (perhaps 10 to 15) and direct a single student or pair of students to sort the shapes into groups. Initially, students can use their own criteria to sort. It is important to include regular and irregular shapes so that they think about where to put some of the "unusual" shapes.

2. Discuss the various sorting categories used: which shapes were included, and why others were excluded. Once the students have completed their sorting, direct them to sort the shapes into some specific categories.

 Direct students to sort by:

- size
- whether a shape has straight or curved sides
- number of sides or corners
- number of sides of equal length
- particular names (e.g. rectangle, oblong, square, triangle, circle, pentagon, and hexagon) or properties (e.g. square corner). Shapes that do not fit any of the known categories can be given informal names by the students. Ensure they are not incorrect geometrically (e.g. a rhombus is not a rectangle).

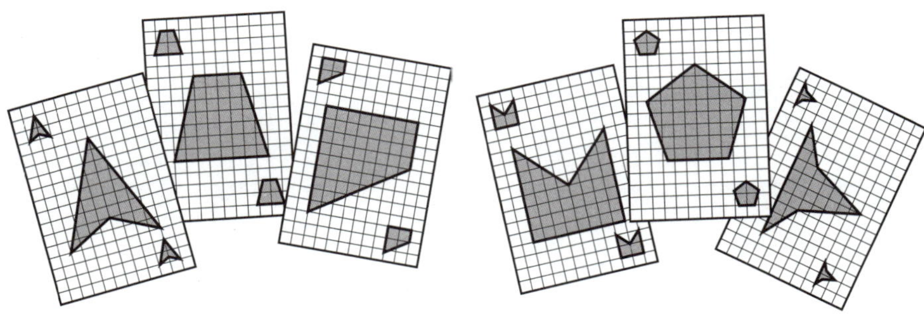

Four-sided shapes   Five-sided shapes

▲ *Students can sort shapes based on number of sides.*

 In addition to the above, students can sort by considering:

- whether the shapes are quadrilaterals
- the presence of 0, 1, 2, 3, or more interior right angles
- whether a shape is regular or irregular
- the presence and number of parallel sides
- triangles by length of side, such as equilateral, isosceles, or scalene

 In addition to the above, students can sort according to:

- whether the shapes are convex or concave

Materials

- Blackline Masters 1 to 5 (pages 59–63)

Did You Know?

An oblong is a rectangle with adjacent sides of different lengths. It is also known as a "non-square rectangle".

Did You Know?

An equilateral triangle has three sides of equal length and three equal angles of 60°.

An isosceles triangle has two sides of equal length and two equal angles.

A scalene triangle has three unequal sides and three unequal angles.

geo Dozens of interesting activities on angles can be found in *All About Angles*.

Materials
- Blackline Masters 1 to 5 (pages 59–63)
- 2 hula hoops

Did You Know?
Venn diagrams are named after John Venn, a British mathematician.

3. Two Or More

Preparation
Copy Blackline Masters 1 to 5 onto tagboard or light card and cut out the cards (or reuse the cards made in Activities 1 and 2).

Activity
1. Place two hula hoops on the ground to represent a Venn diagram.

2. Have the students sort the cards into two groups, using the hoops as an aid. For example, have them place the shapes with four sides inside one hoop and the shapes with right angles in the other. Those shapes that have *both* four sides *and* right angles should be placed in the middle section, where the hoops overlap.

3. Repeat with other properties. Place cards that do not match any category outside the hoops.

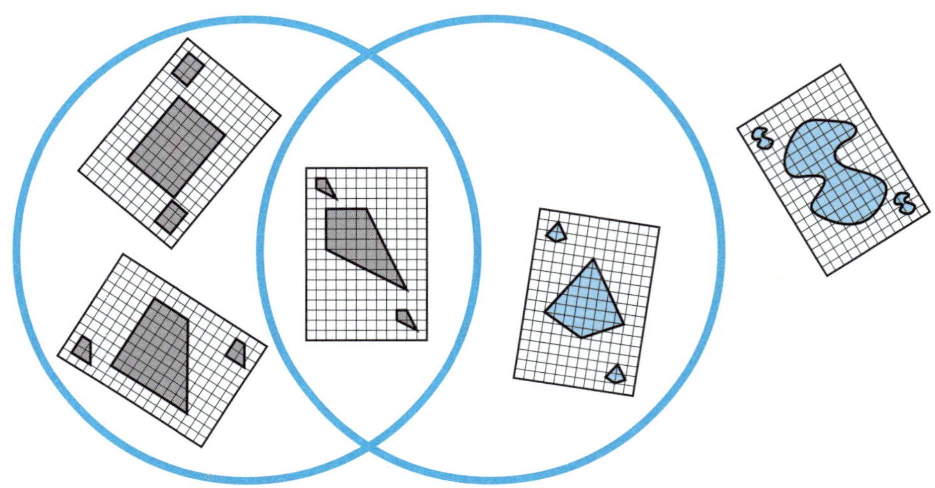

▲ *Have the students sort the cards into those that fit one category, a different category, both categories, or neither.*

Materials
- Blackline Masters 1 to 5 (pages 59–63)

4. Guess My Rule

Preparation
Copy Blackline Masters 1 to 5 on to tagboard or light card and cut out the cards (or reuse the cards made in Activities 1 to 3).

Activity
Ask a student to use one criterion to sort the shapes into two categories. For example, shapes that have four sides and those that do not, or shapes that have at least three sides of equal length and those that do not. The other students have to guess what rule was used to sort the shapes.

5. Ordering Shapes

Preparation

Copy Blackline Masters 1 to 5 onto tagboard or light card and cut out the cards. You need one set of cards for each pair of students.

Materials
- Blackline Masters 1 to 5 (pages 59–63)

Activity

Ask, *How many ways can we order these cards?* Allow the students the opportunity to think of a system of ordering and then place their cards in that order. Lead a discussion about how the students ordered the shapes, for example, ordering by the number of sides or by the number of corners. Invite volunteers who ordered their cards in different ways to show and explain their system.

6. Making a Classification Chart

Preparation

No preparation is required.

Materials
- Tagboard or light card — for each pair or group of students
- Marker pens — for each pair or group of students

Activity

1. An important skill for students of any age is the ability to classify geometric shapes according to their properties. Work with the class to construct a tree diagram that shows a simple classification of mammals.

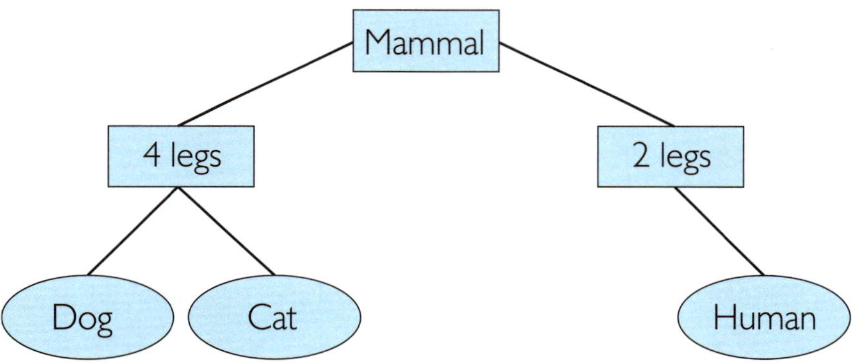

▲ *Constructing a tree diagram can help students examine the relationships between 2D shapes.*

2. Have the students name all the two-dimensional shapes they know. List the shapes on the board, and discuss the properties of each. Activity 2 ("Sorting Shapes"), page 9, provides examples of different properties.

3. Arrange the students into pairs or small groups. Direct each group to classify the shapes using a tree diagram. Have them use the materials to display this information. As a class, discuss each group's classification and any differences of opinion that occurred within and between the groups. (*Note:* There is a classification diagram for two-dimensional shapes in the Introduction to this book on page viii.)

all shapes — representing

Each two-dimensional shape can be represented visually (as a drawing), kinaesthetically (as an object that can be held), and orally (as a description). The following activities are designed to engage students in representing a range of two-dimensional shapes.

■ Materials

- Length of rope with the ends tied together — 1 for each group of students

1. Tied Up

Preparation
No preparation is required.

Activity

1. Discuss aspects of shapes, such as the length of sides and number of corners. Have four students from each group take hold of their rope. Direct them to explore what shapes they can make by changing their positions, and adding or taking away students.

2. As groups make shapes, ask the students to name their shape and describe its attributes.

■ Materials
None

Did You Know?
An oblong is a rectangle with adjacent sides of different lengths. It is also known as a "non-square rectangle".

2. Shape Up

Preparation
No preparation is required.

Activity

1. Have the students use their bodies to make the shapes shown below.

2. Say, *We are going to play "Shape Up". I will call out a shape name and count to ten. If you are not making that shape in a group (or by yourself for a circle) you are out of the game.*

Circle

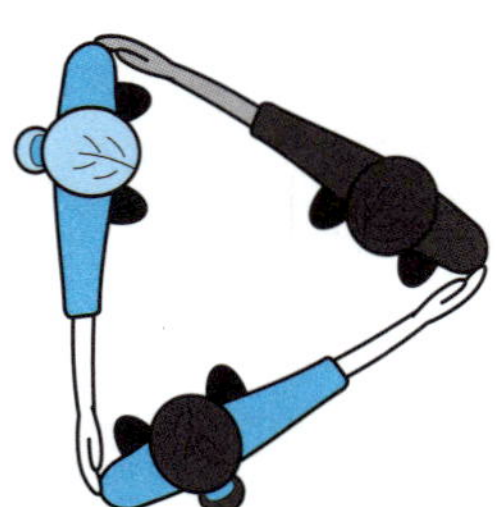

Triangle

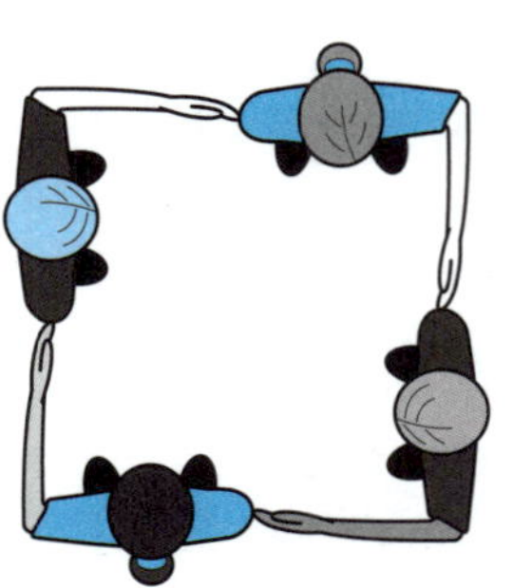

Square

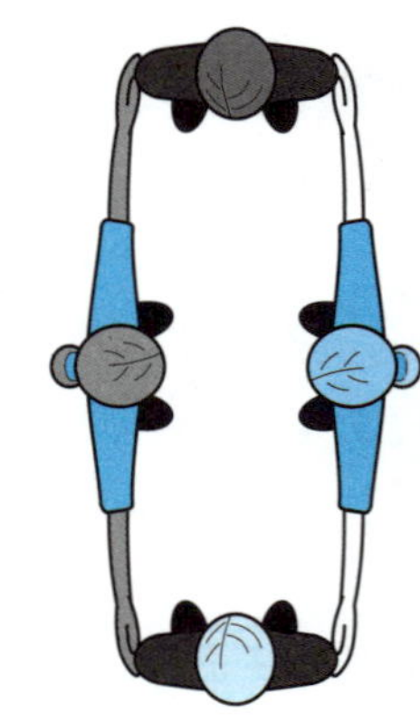

Oblong

▲ *Have the students use their bodies to make simple 2D shapes.*

 Imagery

Preparation
No preparation is required.

Materials
None

Activity
Simple geometric forms can be used to express a variety of emotions. Some of the works of Pablo Picasso and Lyonel Feininger use many simple shapes to convey ideas. Encourage the students to use shapes they have seen inside and outside the classroom to draw, for example, "dizzy" shapes or "happy" shapes. Ask the students to describe the shapes they used and why they used them.

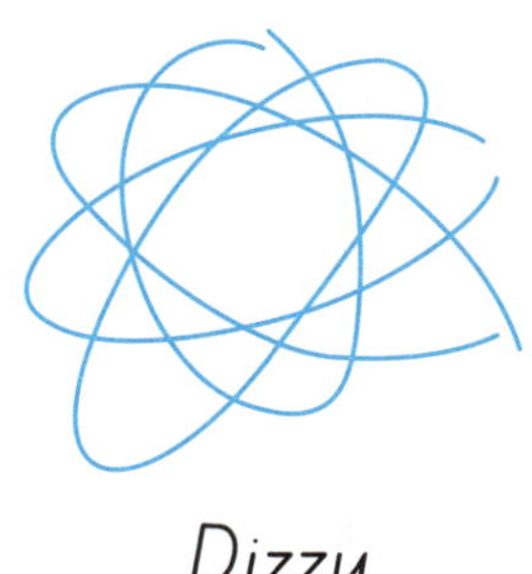

Happy   Dizzy

4. Window Shapes

Materials
• Pattern blocks

Preparation
No preparation is required.

Activity
1. Challenge the students to use multiple copies of a particular pattern block to make a range of "window" shapes. If necessary, display one of the examples below.

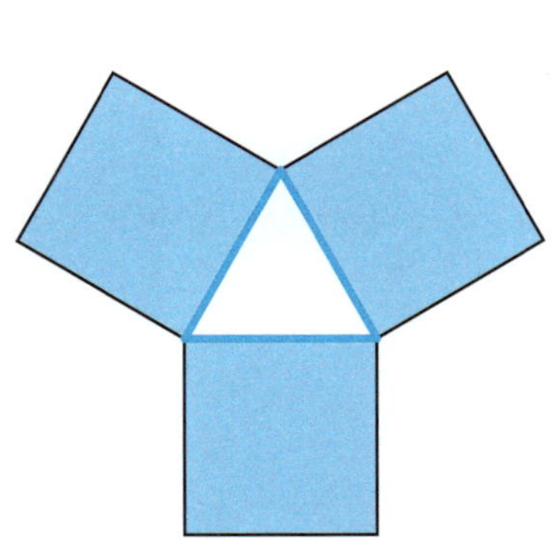

▲ An equilateral triangle made by using squares only.

▲ A concave hexagon that looks like an arrowhead made by using squares only.

▲ A concave decagon that looks like a star made by using rhombuses only.

2. Once the students have created their window shapes, ask them to write a description of the shapes, using a number of everyday and geometric terms. They can then swap descriptions with other students and try to construct their partner's shape.

Materials

- Blackline Masters 7, 8, and 9 (pages 65–67)
- Overhead projector and 3 blank transparency sheets
- Large compass — for the board
- Drawing compass — 1 for each student

5. Compass Constructions

Preparation

Make overhead transparencies of Blackline Masters 7, 8, and 9.

Activity

1. Show the students how to use a compass and share these tips with them:

- *The compass is held at the very top and the needle end is pushed into the paper.*
- *The compass is easiest to turn if the top is rotated between the forefinger and thumb so that the pencil end of the compass turns clockwise (anticlockwise for left-handers).*
- *Lean the top towards the direction of movement so that it makes an angle of about 60° or 70° with the tabletop.*
- *To make sure the compass stays leaning as you turn it, move your entire arm and hand in the same direction as the pencil end of the compass.*

2. Explain the following terms, and draw the diagrams on the board so that the students are familiar with them before completing the rest of this activity:

- *A "point" is a position in space that is marked by a small dot. It can be given a name, for example, point A.*

A

- *An "arc" is part of a circle. It can be described by stating its two end points, for example, arc AB.*

A   **B**

- *A "line segment" is part of a straight line. It can be described by stating its two end points, for example, line segment AB.*

A   **B**

- *The expression "setting a compass" means that the arms of the compass have to be moved so that they are a specific distance apart. For example, if I told you to set your compass to 5 cm, you would need to move the arms of your compass until the tips were 5 cm apart.*

3. Guide the students through constructing the shapes on Blackline Masters 7, 8, and 9, and discuss the following points with them:

- Constructing an equilateral triangle. *How do you make an isosceles triangle using this method? How do you draw a right-angled triangle?*
- Constructing a regular hexagon. *How do you construct a hexagon, starting from a single line segment?*
- Constructing a kite. *What do you think will happen if the distance changes between points A and B?*
- Constructing a parallelogram. *How do you make a rhombus using this method? What do you have to change to make a trapezoid?*
- Constructing a perpendicular line. *What will happen in Step 3 if the compass is set to larger or smaller sizes? Is there a minimum size or maximum size to which the compass can be set?*
- Constructing an oblong. *How do you make a square using this method?*

6. Stained-Glass Windows

Preparation
If necessary, make copies of Blackline Masters 7, 8, and 9 for the students to share.

Activity
A fun way for the students to practice constructing a wide range of regular and irregular polygons (using the techniques on Blackline Masters 7, 8, and 9) is for them to make paper stained-glass windows. Direct the students to draw the shapes lightly on the paper, erase unnecessary construction lines, and color the resulting design. Place the finished designs on a sunlit window to give them the appearance of stained-glass windows.

■ Materials
- Blackline Masters 7, 8, and 9 (pages 65–67)
- Large compass — for the board
- Drawing compass — 1 for each student
- Medium to lightweight paper — 1 sheet for each student
- Color pencils for each student

▲ *These designs can be made using only a ruler, pencils, and compass. See Blackline Masters 7, 8, and 9 for compass instructions.*

geo For more ideas about creating dazzling designs using simple shapes, see *Plane Puzzles, Simple Symmetry,* and *Creative Coverings.*

geo all shapes — combining and dividing

Students have lots of informal experiences with combining and dividing shapes. Using construction toys, doing jigsaw puzzles, and recognizing familiar faces in a crowd are examples students may be familiar with. The following activities focus students' attention on these types of tasks in a more structured way.

■ Materials

- Blackline Master 11 (page 69)
- Blackline Master 12, 13, or 14 (pages 70–72) depending on students' age — 1 sheet for each student
- Scissors — 1 pair for each student

1. Tangrams

Preparation

Copy Blackline Master 11 onto tagboard or light card. You need one set of puzzle pieces for each student.

Activity

1. Direct the students to cut out the tangram pieces and use them to make various shapes from Blackline Masters 12, 13, or 14 depending on their age.

   6–8 Use Puzzles 1 and 2 from Blackline Master 12

   8–10 Use Puzzles 3–5 from Blackline Master 13

   10–12 Use Puzzles 6–8 from Blackline Master 14

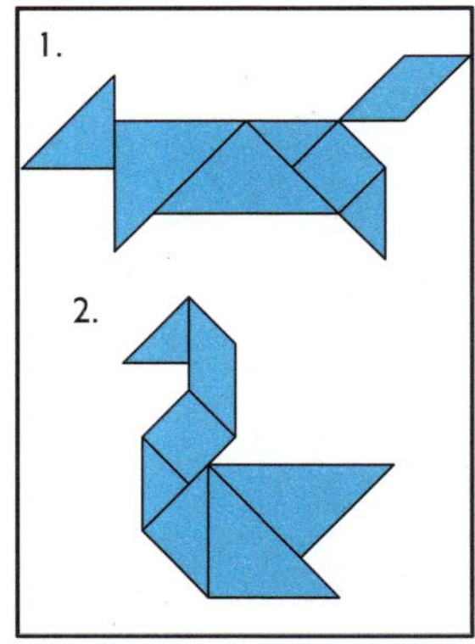 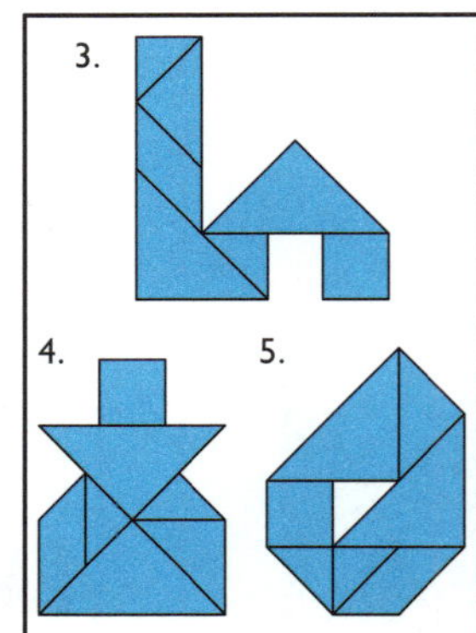 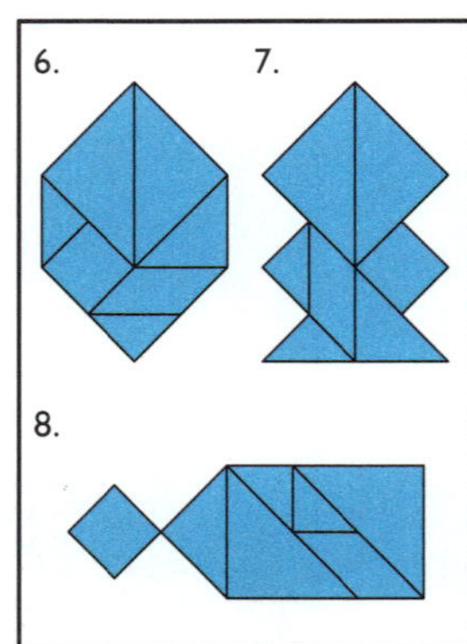

▲ *If some students are having trouble solving the tangram puzzles, you may like to place a piece to help them.*

2. As the students use the puzzle pieces, discuss the pieces and ask questions such as, *Which pieces can be substituted for others?* For example, the three smallest triangles can replace one of the larger triangles. You can also discuss the various properties of the shapes, such as the size of different angles and which shapes have parallel sides.

3. Challenge the students to create their own puzzles for each other to solve. The students may like to create puzzles on a theme, for example animals or people.

2. Polyhexes

Preparation

1. Make one copy of Blackline Master 16 for each student.

2. Trace three pattern blocks onto the overhead transparency sheet so that they make the trihex shown below from Blackline Master 15.

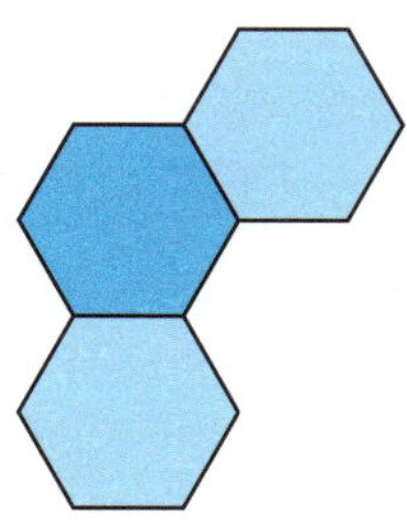

Activity

1. Show the students the trihex you made and explain that a trihex is created by arranging three hexagons so that the sides of the connecting hexagons align perfectly and the vertices touch. Rotations and reflections are considered to be copies.

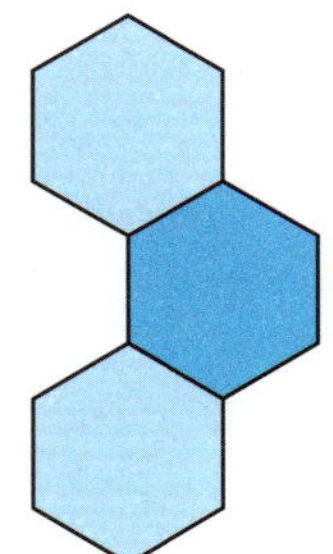 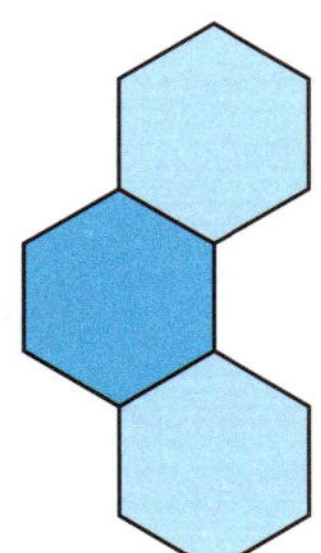 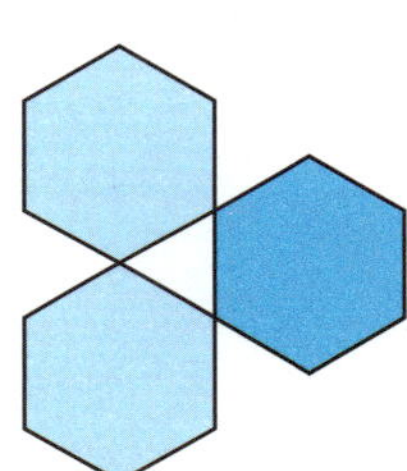

Correct position
The connecting sides align perfectly and the vertices touch.

Not Unique
This shape is a reflection of the one at left.

Incorrect position
The connecting sides do not align perfectly.

Incorrect position
The vertices do not touch.

▲ *Examples and non-examples of trihexes.*

2. Invite the students to investigate how many different shapes they can make from a given number of hexagons. Each student can use the hexagonal grid paper to record their findings.

 Encourage the students to find all of the dihexes, trihexes, and tetrahexes.

 Challenge the students to find all the polyhexes listed above, and all of the pentahexes. Encourage the students to consider how they could create classification groups for the pentahexes. For example, the solutions could be organized according to the length of perimeters, any everyday shapes that they look like, or how many hexagons "in a row" they have.

■ Materials

- Blackline Master 15 (page 73)
- Blackline Master 16 (page 74)
- Hexagonal pattern blocks
- Overhead projector and blank transparency sheet
- Overhead transparency pen

Did You Know?

The term "polyhex" comes from two words: "poly", which means many, and "hex", which is short for hexagon. Other shapes can be used instead of hexagons and are known collectively as "polyforms".

The following prefixes are used to describe the number of shapes in a polyform:

di or do = 2

tri = 3

tetr(a) = 4

pent(a) = 5

hex(a) = 6

geo Other activities on polyforms can be found in *Faces and Frames* and *Simple Symmetry*.

■ Materials

None

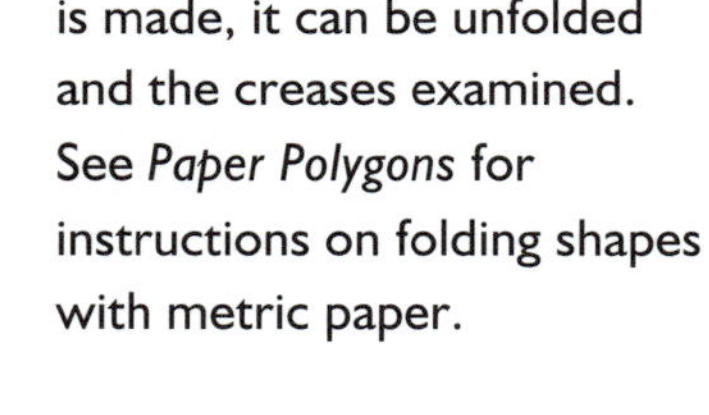

Paper folding is another useful method for finding shapes within shapes. After a shape is made, it can be unfolded and the creases examined. See *Paper Polygons* for instructions on folding shapes with metric paper.

3. Seeing Shapes

Preparation

No preparation is required.

Activity

1. On the board, draw figures similar to the ones below and ask the students questions similar to those provided. These questions help develop student understanding of geometric terms, and reasoning and visual perception abilities. For example, the third figure encourages the students to recall that a square is also a rectangle.

2. The students' answers for problems such as these will often vary. This provides a good opportunity for student-to-student discussion and explanation. Be sure to discuss the students' strategies for answering the questions. Some students may have briefly scanned the picture to see which shapes can be counted, while others may have started with the largest shape, or looked for patterns to make the task easier.

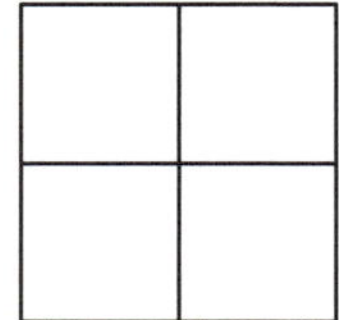 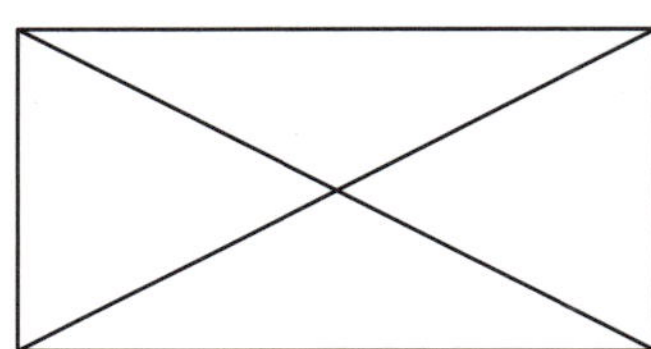 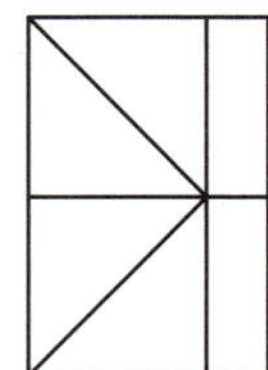

How many squares can you see? (5)   *How many triangles can you see? (8)*   *How many rectangles can you see? (9)*

▲ *These figures can be used to help develop students' understanding of geometric terms, and their reasoning and visual perception abilities.*

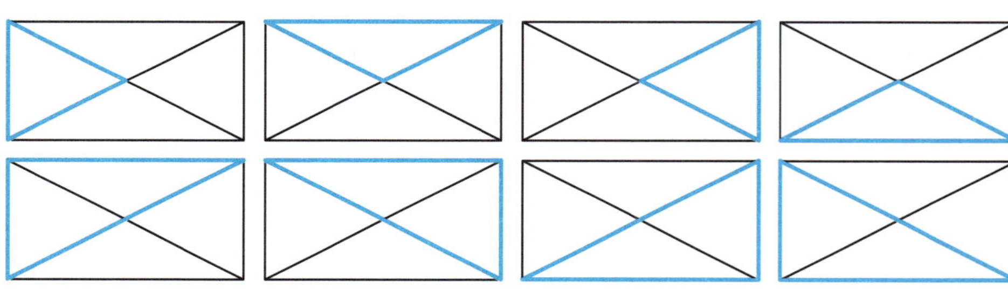

▲ *Show the students how this figure has four individual triangles and four triangles that are made by joining other triangles together.*

■ Materials

• Pattern Blocks

4. Pattern Block Pictures

Preparation

No preparation is required.

Activity

1. A fun activity for all ages, especially for younger students, is making pictures with pattern blocks. The students may use the actual pattern blocks or draw around the pattern blocks and then shade in the shapes. Counters, money, or bottle tops may also be used for tracing.

2. Once a picture is made, direct the students to tally how many of each shape they used.

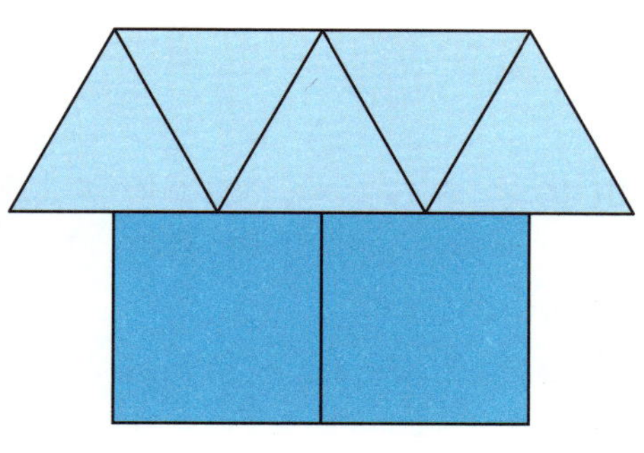

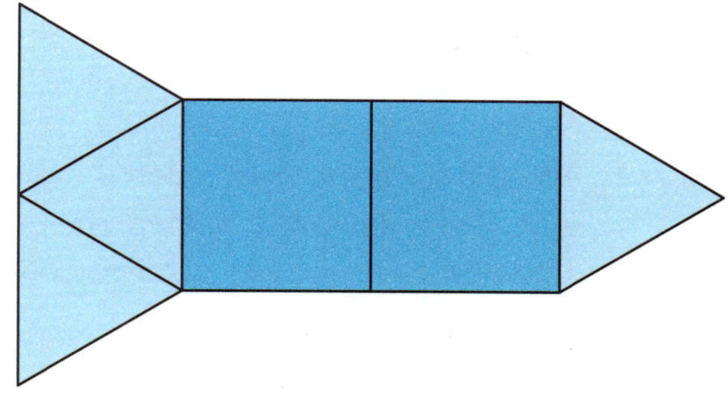

geo **Paper folding** is an enjoyable way for students to make two-dimensional shapes they can use to make pictures. For instructions on how to make polygons out of metric paper, see *Paper Polygons*.

▲ *This house is made up of five triangles and two squares.*

▲ *This rocket is made up of two squares and four triangles.*

5. Circular Jigsaw Puzzles

Preparation
No preparation is required.

Activity
1. Direct the students to draw around the base of their cylinders onto the tagboard.

2. Ask each student to draw a number of lines inside their circle to divide it into no more than ten pieces.

3. Have each student write his or her name on each of the jigsaw pieces, on one side of the tagboard only. The students can then cut out the pieces and swap their puzzle with a partner. Remind the students that the pieces must form a circle.

4. After the students have tried a number of different jigsaw puzzles, discuss which puzzles they found easier or harder, and why. Difficulty may depend on the number of the pieces, the similarity of pieces to one another, and the type of edges each piece has.

5. You might like to have the students try making jigsaws for other shapes (e.g. squares or triangles).

■ Materials
- Tagboard or light card — 2 or 3 sheets for each student
- Large cylinders — at least 10 cm in diameter — 1 for each student
- Scissors — 1 pair for each student

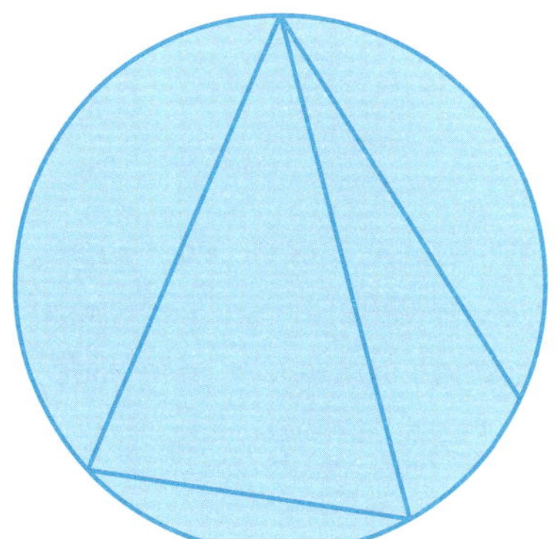 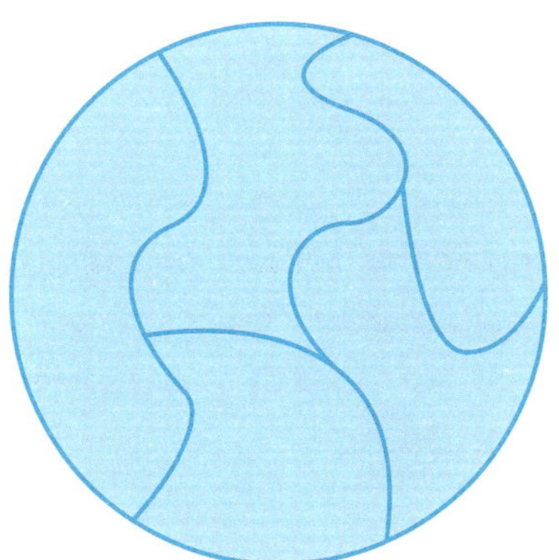 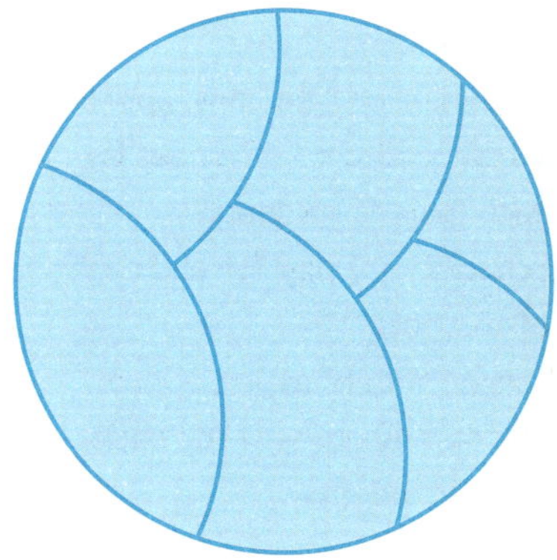

▲ *The type and number of lines the students draw inside their circles contributes to the puzzle difficulty.*

Materials

None

6. Split That Shape

Preparation

No preparation is required.

Activity

1. Ask each student to draw a closed shape with straight sides. Have the students swap their shape with a partner. The partner then draws a straight line segment across the shape to split it into two closed shapes.

2. Direct the students to describe the original shape and then the shapes that resulted from the drawing of the line segment. Challenge the students to see how many different shapes they can make by splitting one particular shape.

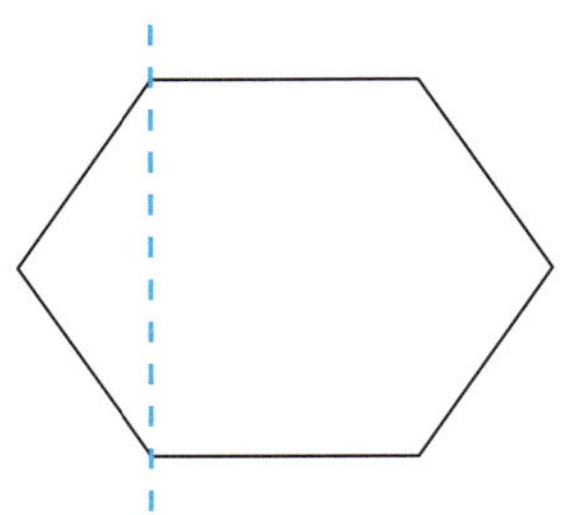

The original shape was a hexagon.
It became a triangle and a pentagon.

▲ *Have students divide polygons into two shapes by drawing one straight line.*

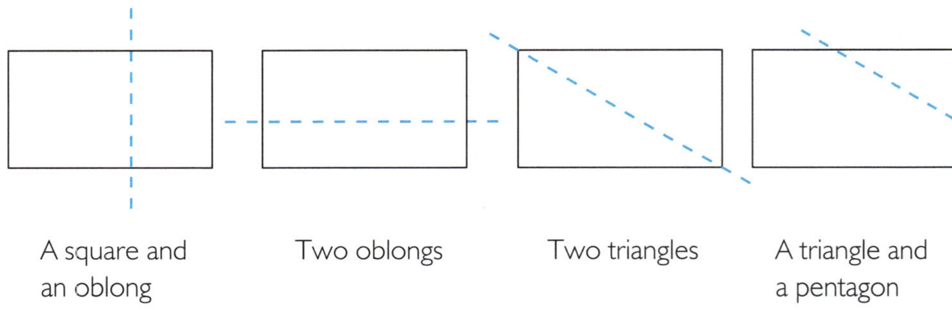

A square and an oblong   Two oblongs   Two triangles   A triangle and a pentagon

▲ *Challenge students to see how many different shapes they can make by dividing one particular polygon.*

Materials

- Tagboard or light card, approx. 10 cm x 15 cm — 1 sheet for each pair of students
- Scissors — 1 pair for each pair of students
- Large sheet of paper, approx. 30 cm x 40 cm — 1 for each pair of students

7. Growing Shapes

Preparation

No preparation is required.

Activity

1. Distribute the tagboard and say, *Draw and cut out a large triangle from your sheet of tagboard. Choose one side of your triangle and write, "base" just above it. Draw a smiley face on the same side of the tagboard. This is the "face" side of the tagboard.*

2. Use one of the student's materials to demonstrate this step of the activity. Say, *The tagboard triangle should be placed face-up on the poster, with the side marked "base" resting horizontally near the top left of the paper. Draw around the entire triangle and label the resulting shape, "triangle". Return the triangle and poster to the student.*

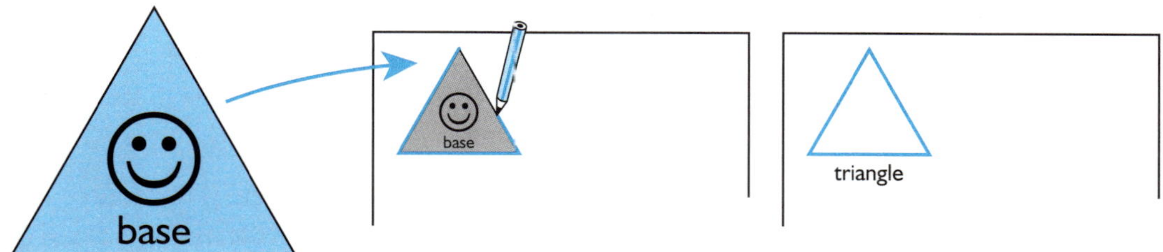

▲ *As the students will be tracing a number of shapes onto the large sheet of paper, have them start in the top left corner.*

3. Say, *Make one straight cut to change your triangle into a quadrilateral.* The students can then draw around the tagboard and label the resulting shape. Have the students continue this sequence so that they make a pentagon, hexagon, heptagon, and octagon, in turn.

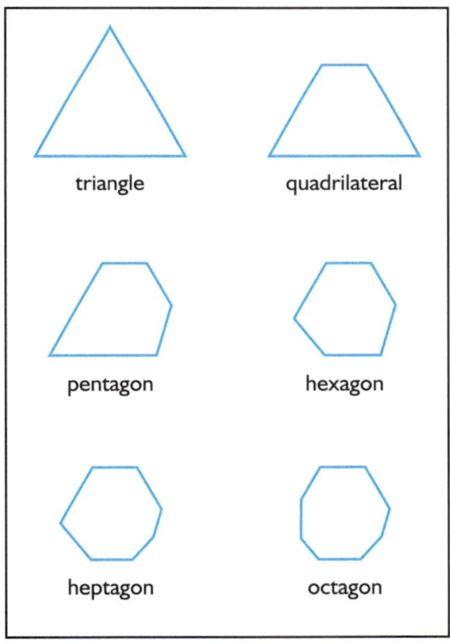

▲ *The students make progressive cuts to make new shapes with more sides.*

4. Afterward, invite the students to show their posters to the class. Discuss the different shapes made by each pair of students. Say, *Even though everybody's shapes look different, they all have some things in common. They all show particular shapes, such as a pentagon. Each time you made a cut, what happened to the number of sides of your template?* (They increased.) *What happened to the area of your template?* (It decreased.) *What was the shape of the piece you cut off to make each new shape?* (A triangle.)

■ Materials

- Tagboard or light card, approx. 10 cm x 15 cm — 1 sheet for each pair of students
- Scissors — 1 pair for each pair of students
- Large sheet of paper, approx. 30 cm x 40 cm — 1 for each pair of students

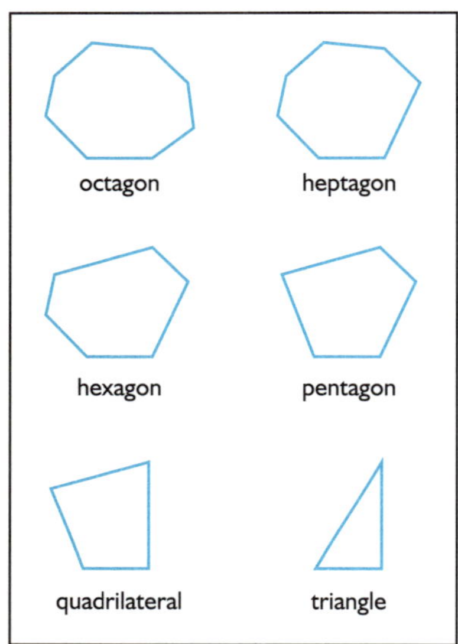

▲ *The students make progressive cuts to make new shapes with fewer sides.*

8. Shrinking Shapes

Preparation

Complete Activity 7, above.

Activity

1. Distribute the tagboard and have each pair of students draw and cut out a large octagon. (*Note:* it doesn't need to be a regular octagon, but must be convex.) Have the students label the template with "base" and a smiley face, draw around it onto the paper, and label the shape.

2. Challenge the students to make one straight cut on the tagboard to reduce the number of sides by one, that is, to make a heptagon. After they draw and label the shape, have them repeat the sequence to make a hexagon, pentagon, quadrilateral, and triangle, in turn.

3. Afterward, discuss the students' posters. Review what the students did in the previous activity, and contrast it to what was done in this one. Say, *In both activities we changed a shape by cutting off a triangle. In one activity, that made the shape have more sides, and in the other, it made it have fewer sides. Why?* Encourage the students to share their thinking.

4. Draw three identical oblongs on the board to aid discussion. Ask, *Where do we cut to change an oblong into a pentagon?* Invite a volunteer to show where to cut. The student should draw a line between two adjoining sides. Call upon another volunteer to draw a line to show how one of the other oblongs can be changed into a triangle. The student should draw a line between opposite corners. Challenge the students to explain why these types of cuts work (A corner-to-corner cut removes two sides and replaces them with one. A cut between two adjoining sides leaves part of two sides and adds another.)

5. Ask, *What will happen if we cut from a corner to a side.* (The shape has the same number of sides but looks different.)

 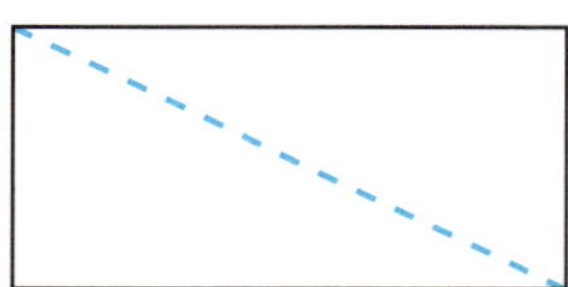 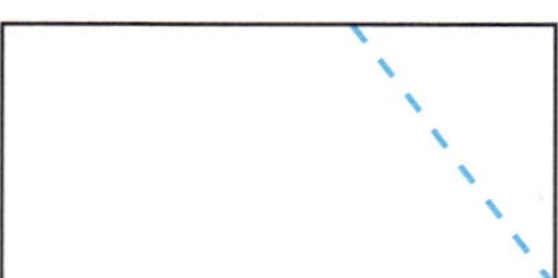

▲ *The students will discover that the position of a cut will affect whether the new shape has more, less, or the same number of sides than the original shape.*

■ Materials
- Pattern blocks
- Plastic counters

9. Overlapping Shapes

Preparation

No preparation is required.

Activity

1. Students can make many interesting designs by overlapping pattern block and counters outlines. They might like to erase unnecessary lines or keep the lines and shade the resulting regions.

2. Discuss the students' designs and ask them to name any new shapes that were formed by intersecting lines. For example, the design shown below was made from equilateral triangles and a square. The intersecting lines also form trapezoids, kites, and smaller equilateral triangles.

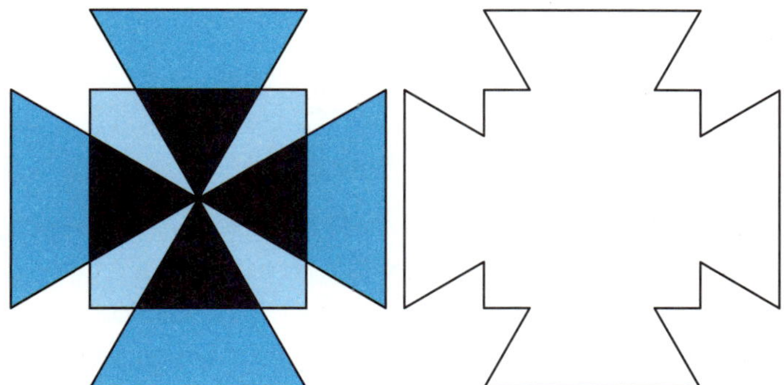

▲ *Students can either shade the regions on their design or erase the internal lines.*

10. Triangles In Shapes

Preparation
No preparation is required.

Materials
None

Activity

1. Direct the students to draw the following table. Omit the numbers (these are the answers).

Shape	Number of Sides	Number of non-intersecting diagonal lines	Number of triangles formed
Triangle	3	0	1
Quadrilateral	4	1	2
Pentagon	5	2	3
Hexagon	6	3	4
Heptagon	7	4	5
Octagon	8	5	6

2. Have the students draw each of the polygons listed in the table. The shapes do not need to be regular.

3. Say, *In each shape draw non-intersecting diagonal lines from one corner to another, to make as many triangles as you can. Record the results in your table.*

4. Discuss the pattern that forms and encourage the students to state a rule that describes the relationship. For example, *The number of diagonals is three less than the number of sides and the number of triangles is two less than the number of sides.* Challenge the students to predict the results for a 20-sided shape (20 sides, 17 diagonals, 18 triangles) and a 100-sided shape (100 sides, 97 diagonals, 98 triangles). Ask questions such as, *How many sides would a shape have if 14 triangles could be made within it?* (16.)

■ Materials

- Blackline Master 15 (page 73)
- Blackline Master 17 (page 75)
- Paper and drawing compass
  — for demonstration
- Paper — 1 sheet for each
  student
- Drawing compass — 1 for
  each student
- Scissors — 1 pair for each pair
  of students
- Large sheet of paper, approx.
  30 cm x 40 cm — 1 for each
  pair of students
- Glue — for each pair of
  students
- Adhesive tape

Did You Know?

The term "polyquad" comes
from two words: *poly*, which
means many, and quadrant.
Other shapes can be used
instead of quadrants and
are known collectively
as "polyforms".

11. Polyquads

Preparation

1. Draw a circle on a blank sheet of paper, cut it out, and then cut it into quadrants. Arrange the quadrants to form one of the tetraquads (apart from the circle) shown on Blackline Master 15.

2. Make one copy of Blackline Master 17 for each student.

Activity

1. Show the students the tetraquad you made and explain that a tetraquad is created by arranging four quadrants so that the sides (i.e. the radii) of the connecting quadrants align perfectly and the vertices touch. Arcs cannot be joined to each other or to straight sides. Rotations and reflections are considered to be copies.

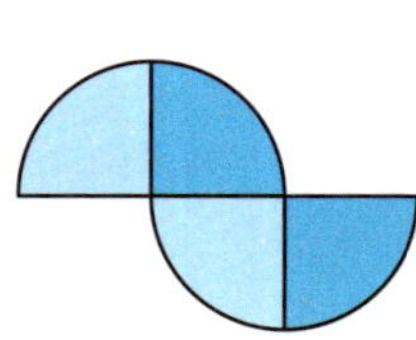 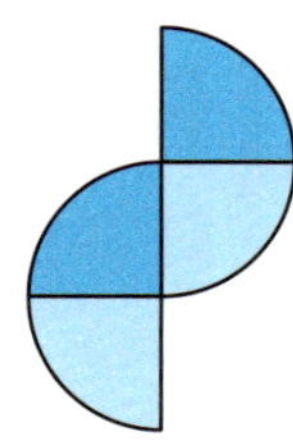  

Correct position
The connecting sides align perfectly and the vertices touch.

Not Unique
This shape is a rotation of the one at left.

Incorrect position
The connecting sides do not align perfectly.

Incorrect position
The vertices do not touch.

▲ *Examples and non-examples of tetraquads.*

2. Challenge the students to find out how many different shapes they can make from a given number of quadrants. Have each student draw two circles of equal size and cut them into quadrants. The students can then work with a partner to investigate how the quadrants can be arranged to form diquads, triquads, tetraquads, and pentaquads. (See Blackline Master 15.)

3. The students can use the circular grid paper (Blackline Master 17) to record their investigations. Explain that it is much easier to find all the designs if they work systematically. To explore tetraquads, for example, they might find it easier to begin with one of the three triquad designs and then add another quadrant to find all the variations. They could then choose another triquad and repeat the process.

4. Discuss the results with the class. Ask for suggestions as to how the results could be organized. One option is to create a tree diagram linking a particular diquad to related triquads, tetraquads, and pentaquads. In such a system, the students will find that the more the number of quadrants increases, the more individual designs can fit into a number of different categories. Encourage the pairs of students to cut out and arrange their polyquads into a system and display their results as posters.

12. Making Tangram Pieces

Preparation

Make an overhead transparency (OHT) of Blackline Master 11.

Activity

1. Direct the students to make their own set of tangram pieces using the sheet of square paper. Show them the OHT of Blackline Master 11 as a guide.

2. Some students may need to experiment with an actual set of tangram pieces to discover the relationships that will help them complete the task. (Check they do not just trace the pieces.) A solution is provided below, but guide students to discover their own methods by encouraging them to look at where corners touch other shapes after folding. They are located at midpoints on the sides of other shapes. This means that those corners can be found by folding the other shapes in half.

Materials

- Blackline Master 11 (page 69)
- Overhead projector and blank transparency sheet
- Square paper — 1 sheet for each student

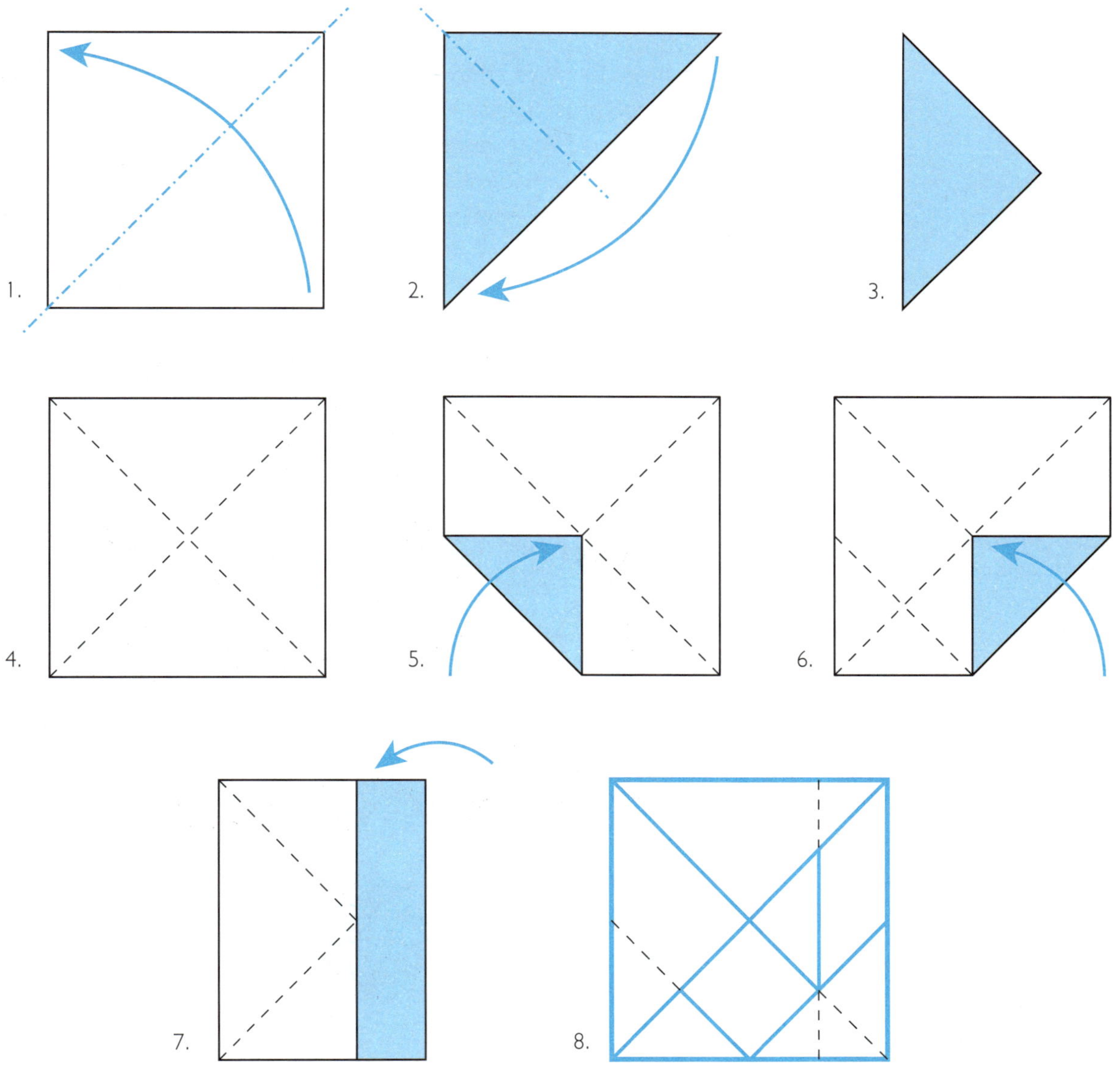

▲ *One solution for creating a tangram set from a square sheet of paper.*

Materials

- Blackline Master 18 (page 76)
- Overhead projector and blank transparency sheet

Did You Know?

The word "squangles" is a combination of "squares" and "triangles".

Did You Know?

A shape is said to have "line" or "reflective symmetry" if half of the shape is a mirror image of its other half. If a shape looks the same after being rotated through a fraction of a turn, then it is said to have "rotational symmetry".

Shapes that are used to cover a plane to make a repeating pattern without any gaps are said to "tessellate". The result is called a "tessellation".

Many interesting activities on symmetry and tessellations can be found in *Simple Symmetry* and *Creative Coverings*.

13. Squangles

Preparation

1. Make an overhead transparency (OHT) of Blackline Master 18.

2. Make one copy of Blackline Master 18 for each student.

Activity

1. Show the students the OHT of Blackline Master 18 and describe how the shapes in the tessellation can be joined to form new shapes. Use the example of a pentagon being formed by a square and a triangle.

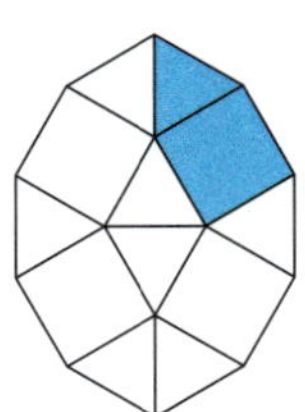

2. Challenge the students to lightly shade shapes on Blackline Master 18 to show examples of polygons, ranging from those with 3 sides to those with 10 or more. The students should write inside each composite shape the number of sides it has. They may like to shade shapes that show reflective or rotational symmetry. Some examples that have reflective symmetry are shown below.

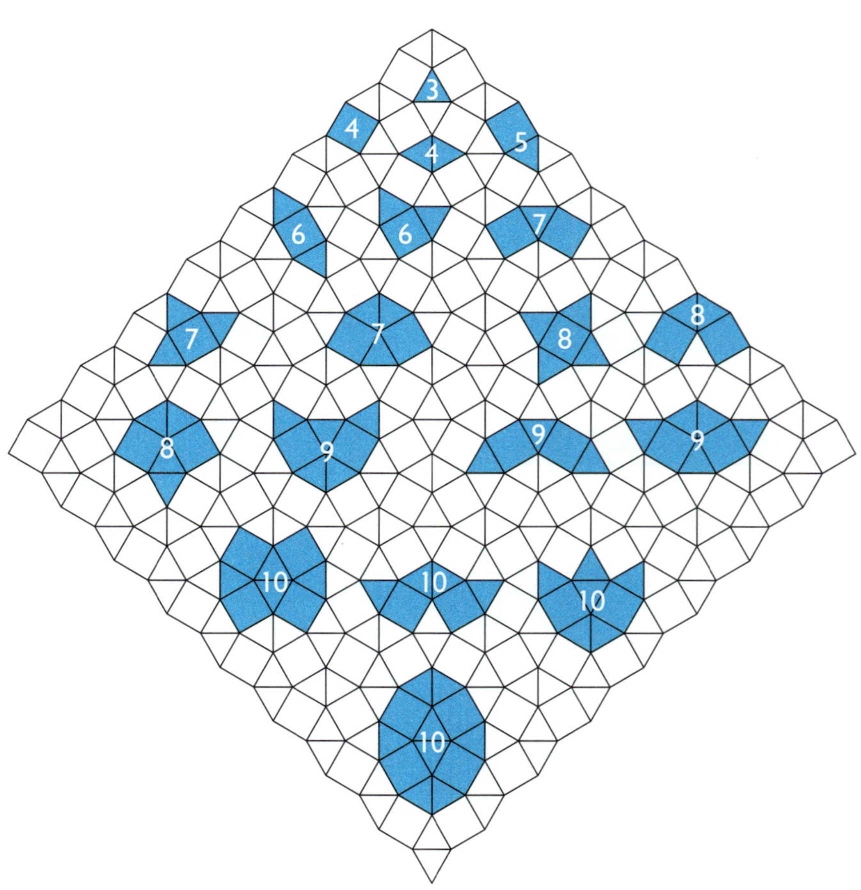

▲ *Challenge the students to find a range of different polygons by combining squares and triangles.*

 *geo* all shapes — copying, enlarging, and reducing

Reproducing shapes involves thinking about position and proportion. To make a polygon keep its shape regardless of size, the proportions of the sides to one another must be exactly as they were in the original. The angles that the sides show must also be exactly the same as they were before. Changing the proportions of the sides changes the size of the angles. Similarly, changing the angle sizes may change the proportions of the sides to one another. The following activities engage students in thinking about these ideas.

1. Big Blocks

Preparation

No preparation is required.

Activity

1. Have the students establish the fact that the shortest sides of all the pattern blocks are the same length.

2. Challenge the students to use one type of pattern block to make a copy of the shape that has sides twice as long as the original. The solutions are shown below. The students will discover that a larger copy of the hexagon cannot be made using the hexagons alone.

■ Materials

- Pattern blocks — 1 set for the whole class

Did You Know?

In North America, the term "trapezoid" is used to describe a quadrilateral that has only two parallel sides. In Australia, Europe, and elsewhere, the same shape is called a "trapezium".

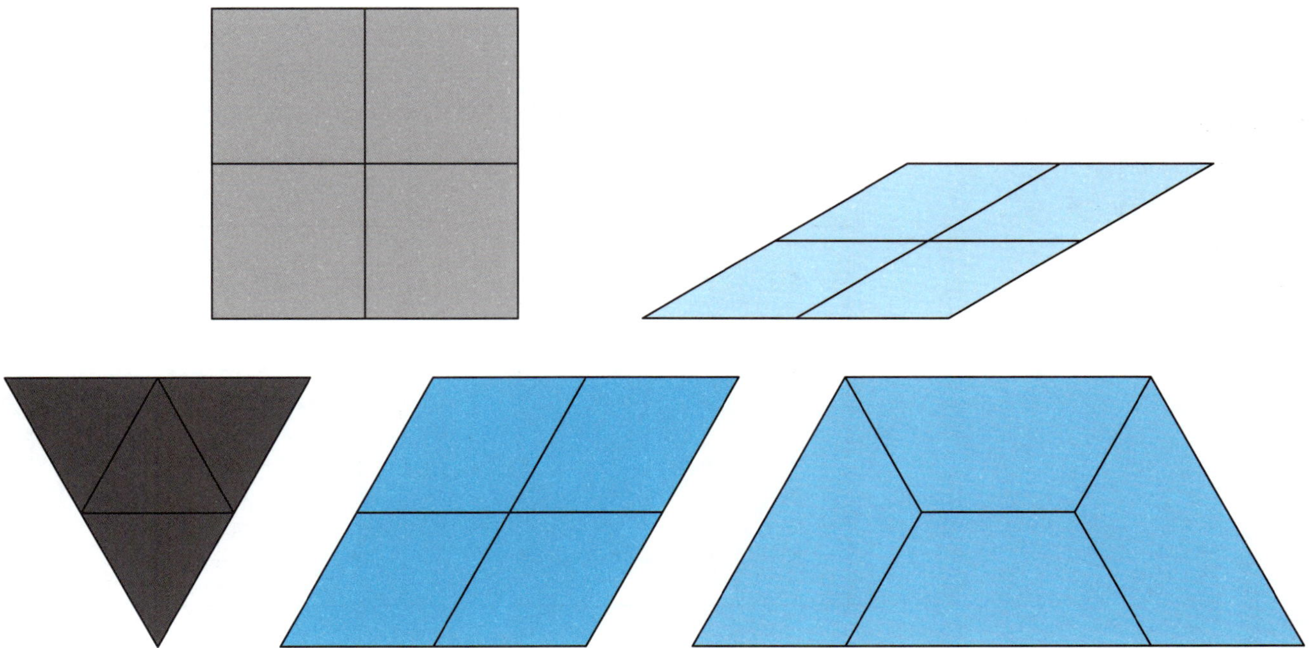

▲ *Challenge the students to create larger versions of pattern blocks by combining smaller ones.*

A shape that is created by joining together other smaller shapes is called a "compound" shape. See *Paper Polygons* for many activities that involve compound shapes.

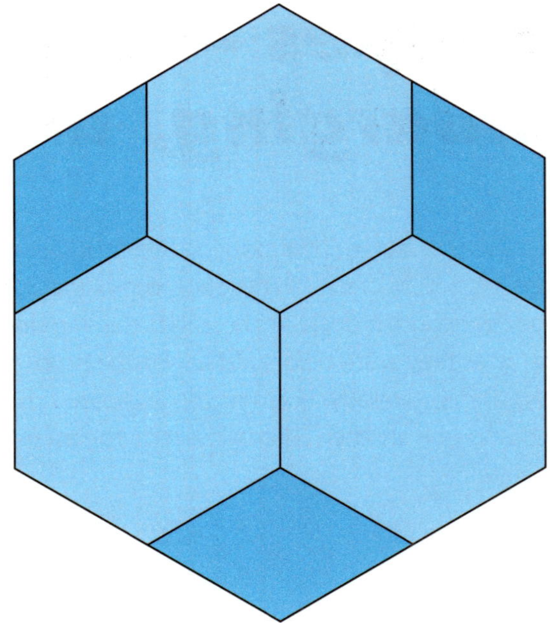

▲ *One solution for creating a larger hexagon using pattern blocks.*

Materials
- Geoboards and rubber bands for each student

Reflective symmetry involves making mirror images of shapes. For dozens of fascinating activities on reflective symmetry, see *Simple Symmetry*.

2. Copy It

Preparation
No preparation is required.

Activity
1. Have one student in each pair make a simple shape on a geoboard. Ask that student to show the shape to his or her partner. The partner has to copy it onto their geoboard.

2. Once the students become familiar with the activity, encourage them to make more complex shapes. Challenge the students to copy their partner's shape back to front, like a mirror image.

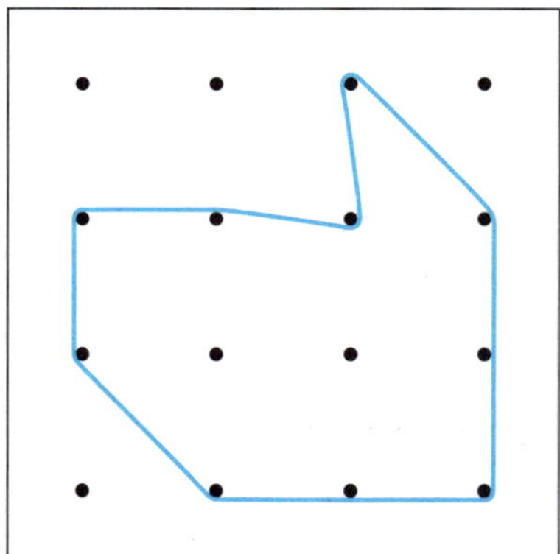 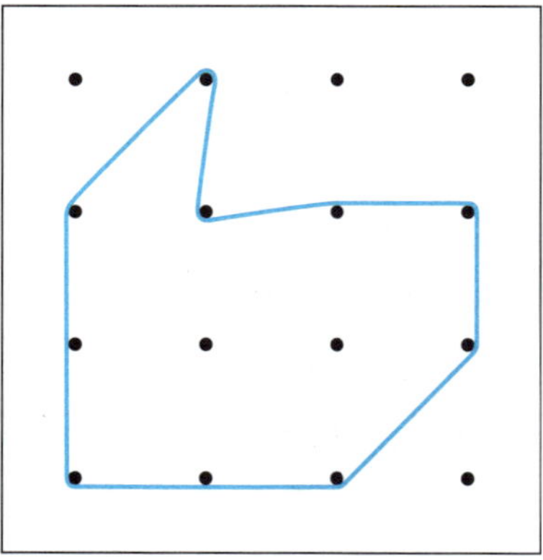

▲ *Students may find creating a mirror image challenging.*

3. Making It Big

Preparation
No preparation is required.

Activity
Have one student in each pair make a small, simple shape on a geoboard. The student's partner then uses his or her geoboard to make a larger copy of the same shape. For example, if a square has a side of 2 units, a larger copy with a side of 3 units could be made. Or, if space allows, the side lengths could be doubled.

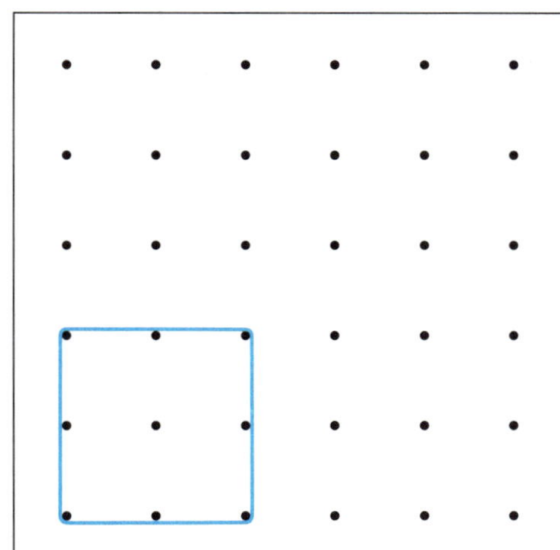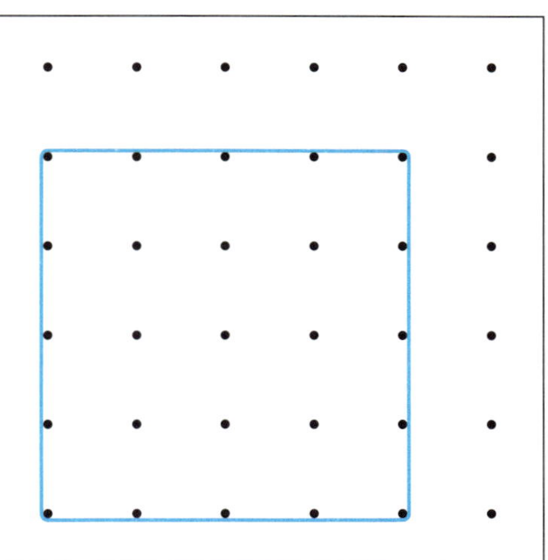

▲ *Challenge the students to make enlarged versions of shapes.*

Materials
- Geoboards and rubber bands for each student

4. Remember It

Preparation
No preparation is required.

Activity
1. Use the overhead pattern blocks to project a design that students have to copy with their own pattern blocks. Begin with designs that are symmetrical and use only a few pieces before trying more unusual shapes.

2. Once the students can copy the designs accurately, direct them to memorize a design you make. Show the design for 5 to 10 seconds before covering it up. The students can then attempt to recreate the design.

3. Once the students feel they have completed all they can, uncover and display the design for another 5 to 10 seconds. Cover the design and allow the students to correct their designs if necessary.

4. Discuss the strategies that the students used, such as memorizing small combinations of shapes. The students could also be asked to draw some designs freehand.

5. A harder alternative is to make your own overhead blocks from overhead transparencies. Leave them clear so that the emphasis is on shape, not color.

Materials
- Overhead projector
- Overhead pattern blocks (see note)
- Pattern blocks — 1 set for the whole class

Teaching Note
Overhead pattern blocks are transparent pattern blocks that can be used on overhead projectors. If you do not have access to overhead pattern blocks, you may like to use ordinary pattern blocks and do this activity with half the class at a time, sitting on the floor, with the blocks in the middle.

Materials

- Blackline Master 19 (page 77)

Did You Know?

A horizontal line or surface is completely level.

A vertical line or surface forms a 90° angle with a horizontal line or surface.

A line or surface which is neither vertical nor horizontal is called "oblique".

geo Changing the dimensions of shapes is an example of "topology". For dozens of fascinating activities involving topology, see *Twists and Turns*.

5. Stretch It

Preparation

Make one copy of Blackline Master 19 for each student.

Activity

1. Have the students draw a small shape (no wider than 4 x 4 units) on the grid squares at the top left of Blackline Master 19. Explain that each of the other grids will change the shape in some way when it is copied onto them. Remind the students that if one side of a shape is 3 units long on the original grid, then each time it is copied it must also be 3 units long.

2. As each student copies his or her shape, discuss how the shape changed. Encourage the students to describe the effects on horizontal and vertical dimensions. For example, one type of grid reduces the length of horizontal measurements, while another changes vertical lines to oblique lines.

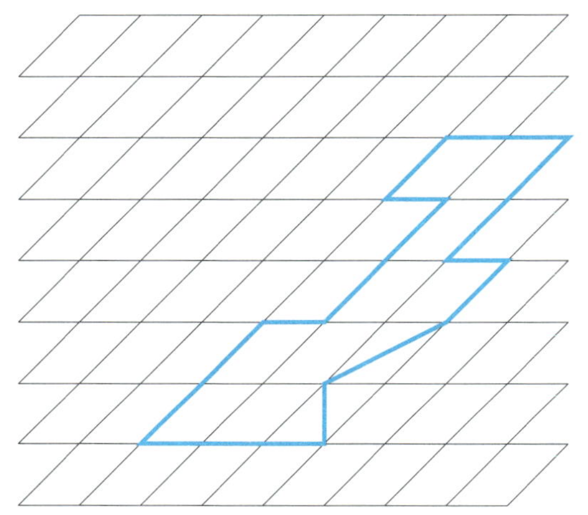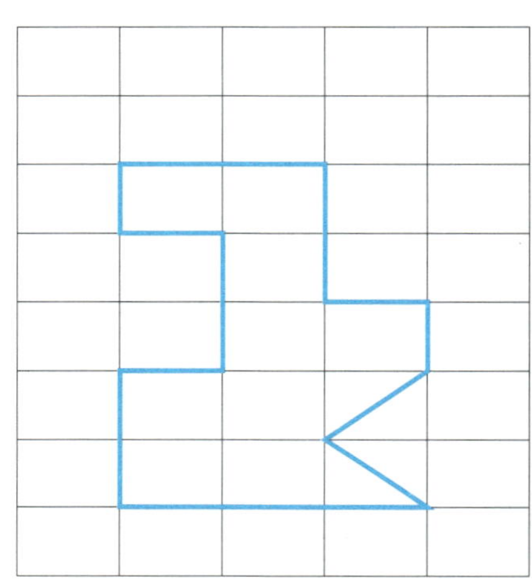

▲ As the students copy a shape onto different types of grids, talk about how the shape changes.

Materials

- Drawing compass — 1 for each student
- Large compass — for the board

Did You Know?

The word "concentric" comes from the Latin roots, *com-*, meaning "together" and *centrum*, which means "center".

6. Concentric Circles

Preparation

No preparation is required.

Activity

1. Direct the students to draw a circle of any size while you draw one on the board. Ask, *How can we draw another circle of a different size so that the two circles intersect? Where will the center of the new circle be?* Invite a volunteer to indicate a suitable point on your circle.

2. Ask, *If we wanted to make sure that the circles didn't intersect, where could we place the center of the second circle?* Discuss the students' suggestions.

3. Have the students set their compasses to a different size from their original circles. Say, *If we want to make sure that two circles never intersect, we can use the same center for each circle.* Have the students draw a number of circles of different diameters, using the same central point for each one. Explain that the circles they drew can be described as being "concentric", a word that means the shapes have the same central point.

▲ *The students will find that concentric circles never intersect.*

7. Resizing from the Middle

Preparation
No preparation is required.

Activity

1. Demonstrate the method, as shown in the figure below, of enlarging a shape to produce a mathematically similar shape. The students will find that using a compass to copy and mark the necessary distances is easier than using a ruler.

■ Materials
- Drawing compasses — 1 for each student
- Large compass — for the board

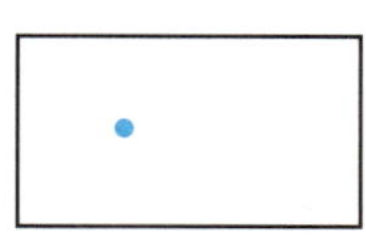

a. Draw a point inside the shape.

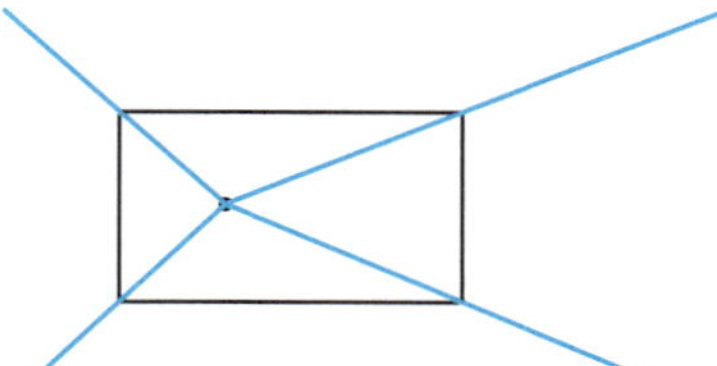

b. Draw guide lines from this point so that they pass through each of the shape's vertices.

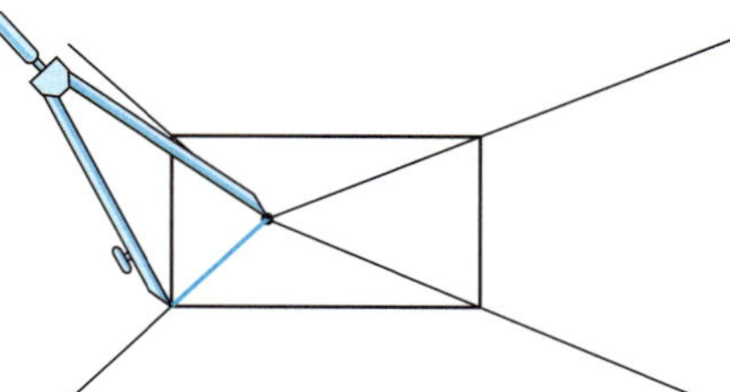

c. Set the compass to the distance between the point and one of the vertices.

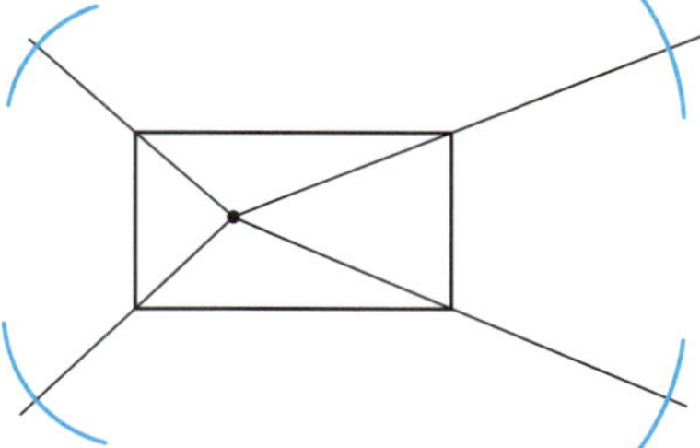

d. From the vertex, draw an arc to intersect the guide line outside the shape.

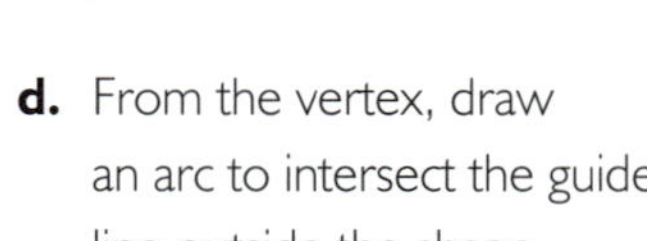

e. Repeat the two previous steps with each vertex and corresponding guide line.

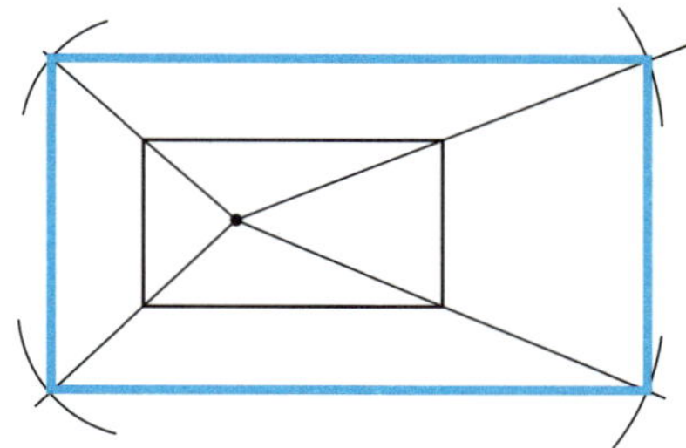

f. Join all four intersecting points together to complete the new shape.

▲ *Demonstrate these steps for enlarging a shape.*

2. Say, *Measure the perimeter of both of the shapes. What do you notice?* (The perimeter of the larger shape is twice that of the original.) Highlight the fact that the shapes are also concentric.

3. Although it is simple to construct a shape that is double the perimeter of another shape, more steps are involved to produce an enlargement that has, for example, a perimeter one and a half times the size of the original. Challenge the students to construct such a shape. If they need help, guide them to see that all of the measurements along the guide lines outside the shape will have to be half of the distances from the inside point to the vertices.

geo Other shapes besides circles can also be concentric. Concentric polygons can be easily made by folding A-sized metric paper. See *Paper Polygons* for instructions.

■ Materials

- Blackline Master 10 (page 68)
- Overhead projector and a blank transparency sheet
- Large compass — for the board
- Drawing compasses — 1 for each student

Teaching Note

Refer to "Compass Constructions" on p.14 for tips on using a compass.

Did You Know?

An arc is a part of any curve and is often used to mean part of a circle.

The point where two sides of a polygon meet is called a "vertex".

Shapes that are exactly the same size and shape are said to be "congruent".

Shapes that look exactly the same, apart from their size, are said to be "similar".

Did You Know?

In everyday use, a "line segment" is referred to as a "line". Mathematically, however, a line can be thought of as extending infinitely into space, while a line segment is just a part of that infinite line. You may like to model the correct terminology to your students, but it is not overly important that they remember it.

8. Drawing Copies of Shapes

Preparation

Make an overhead transparency (OHT) of Blackline Master 10.

Activity

1. Display the OHT of Blackline Master 10 and demonstrate the steps for drawing a congruent copy of a shape. Emphasize that the distance between the tips of the compass arms is the same distance between two vertices of the shape.

2. Have the students draw and copy a number of different shapes. Once they appear comfortable using the technique, ask them to draw a triangle with sides that are 3, 4, and 5 units long. Remind them to draw an initial line, mark off the vertices for the first side, and then draw intersecting arcs from these points to find the remaining vertex.

a. Draw a line 3 units long and mark points A and B. These will be two of the triangle's vertices.

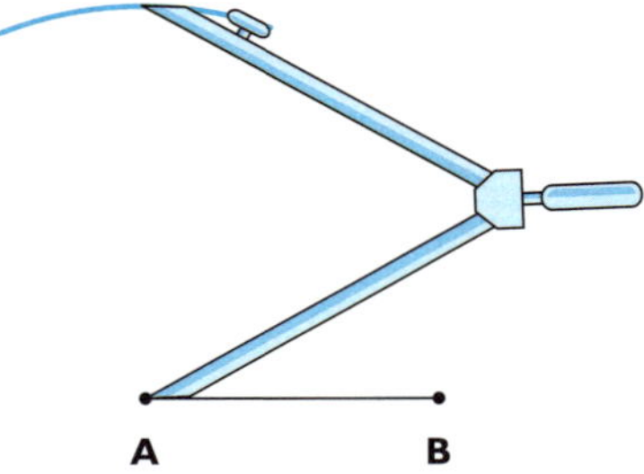

b. Set the compass to 4 units long. Mark this distance from point A.

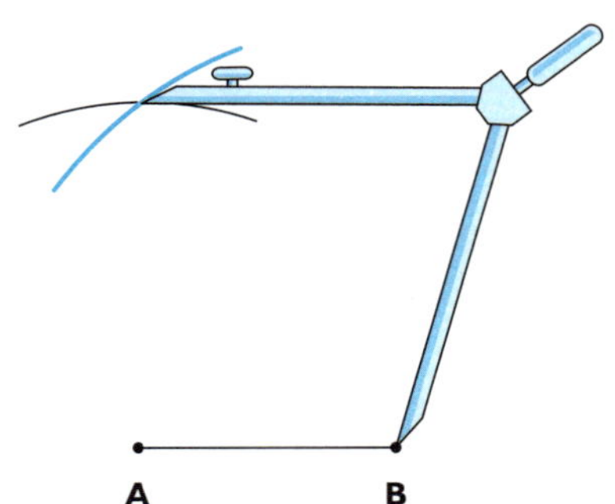

c. Set the compass to 5 units long. Mark this distance from point B.

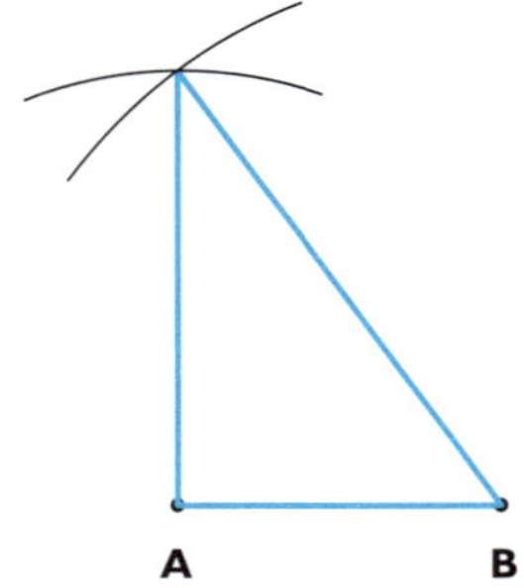

d. Where the arcs intersect is the third vertex. Draw lines from points A and B to this vertex to complete the triangle.

▲ *Show the students the technique for drawing a triangle when given the dimensions only. This one has sides of 3, 4, and 5 units.*

3. Challenge the students to make mathematically similar copies of one of their shapes. For example, they can halve the lengths of each side. In using Blackline Master 10, for example, the students must also halve the diagonal distances from A to C and from B to D.

Many students may recognize a circle as a particular shape but not realize the relationship between the center of the circle and all the points on the circumference. The following activities provide students with opportunities to explore the characteristics of circles, parts of circles, and ellipses.

1. What Makes a Circle a Circle?

Preparation
Cut the bottom off the drink bottle.

Activity

1. Take the students to an open, grassy area. Hammer the stake into the ground and make a loop at one end of the string, to fit loosely over the stake. Attach the other end of the string to the neck of the bottle and pull the string taut.

2. Ask a volunteer to hold the bottle upside down. Pour the sand into the bottle and direct the student to walk around the stake so that the string stays taut. The sand pouring out of the bottle will form a circle.

3. Direct a number of students to stand at various points on the circle. Ask, *What do you think the stake represents?* (The middle of the circle.) *How far away from the stake is each student on the circle?* If necessary, have a volunteer use the string to check that the students on the circle are all the same distance away from the center.

4. Create other circles by shortening or lengthening the string, and repeating the process above.

Materials
- Wooden stake
- Hammer
- 3–6 ft (1–2 m) length of string
- Empty plastic drink bottle
- Bucket of sand

2. My Own Circle

Preparation
No preparation is required.

Activity

1. Arrange the students into small groups. Have the students flatten the clay to make a thin slab of about 1 cm thick.

2. Ask each student to select a twig and place one part of the fork on the clay to mark the center of a circle. By laying the other part of the fork on the clay, students will be able to draw a circle by rotating the top of the twig between their forefinger and thumb.

Materials
- Large ball of clay and clay mats — 1 set for each student
- Sturdy twigs that are forked at one end — a range of sizes for each group of students

Teaching Note
If the students cannot find strong enough twigs, or clay is not available, damp sand may be used as an alternative to clay.

8–10

■ Materials

- Circular containers of various diameters — 1 for each student
- Markers — 1 for each student
- Several sheets of paper in different colors — 1 set for each student
- Scissors — 1 pair for each student
- Ruler — 1 for each student
- Sheet of thick cardboard and a drawing pin — 1 set for each student

Did You Know?

Mathematically speaking, only the circumference of a circle is a circle. The area inside is called a disc, although sometimes the circle and area inside together are called by this name. When you "cut out a circle" drawn on paper, you are actually cutting around the circle, complete with the disc inside.

---

3. Encourage the students to draw overlapping circles in the clay. Also, have them use twigs with forks different distances apart and use the same center as a previously made circle.

4. Choose two twigs with forks different distances apart. Ask, *What do you know about the circles that can be made with these twigs?* The students may say that the circles will be of different sizes, and that the larger twig will make the larger circle.

3. Folding a Circle

Preparation

No preparation is required.

Activity

1. Have each student place a container on a sheet of paper and draw around its base. Ask, *How could you work out the middle point, or center, of your circle?* Discuss the students' suggestions, then direct them to estimate and mark the center of their circle.

2. Direct the students to cut out their circles. Instruct them to fold the circle in half, unfold it, then fold it in half from a different point on the circumference, and unfold it.

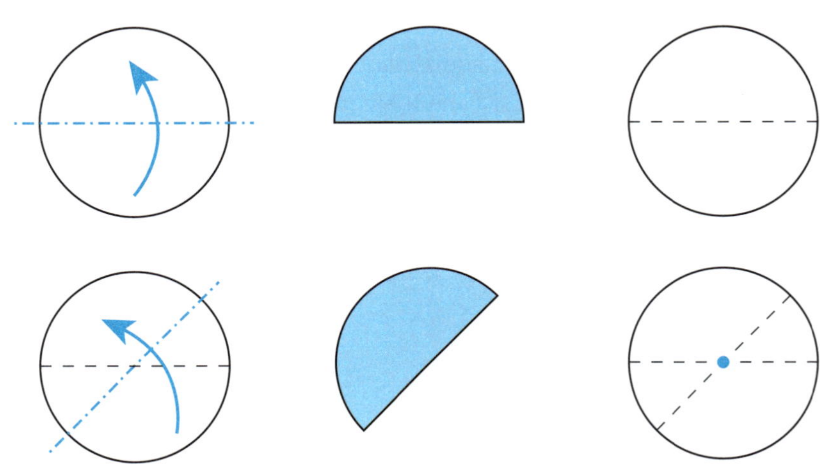

▲ *Have the students fold their circular pieces of paper in half twice to find the center.*

3. Say, *Where the creases cross over one another is the center of your circle.* Ask, *Can you prove that point is the center?* Bring out the idea that the distances from the center to all points on the circumference will be the same. The students can use a ruler to measure the distances and check that the intersecting creases do, in fact, mark the center of the circle.

4. Invite the students to draw other circles of different sizes, cut them out and find the centers. They can then align the centers and attach them to the cardboard for display.

4. Semicircles

Preparation

Make and cut out a demonstration circle using the large sheet of paper and drawing compass.

Activity

1. Ask each student to draw a circle on a sheet of paper, and then cut out the circle.

2. Direct the students to fold their circular piece of paper into quarters. Have them unfold their paper and identify the center of the circle. If necessary, introduce the terms "diameter" and "radius", and have the students identify examples of each on their shape.

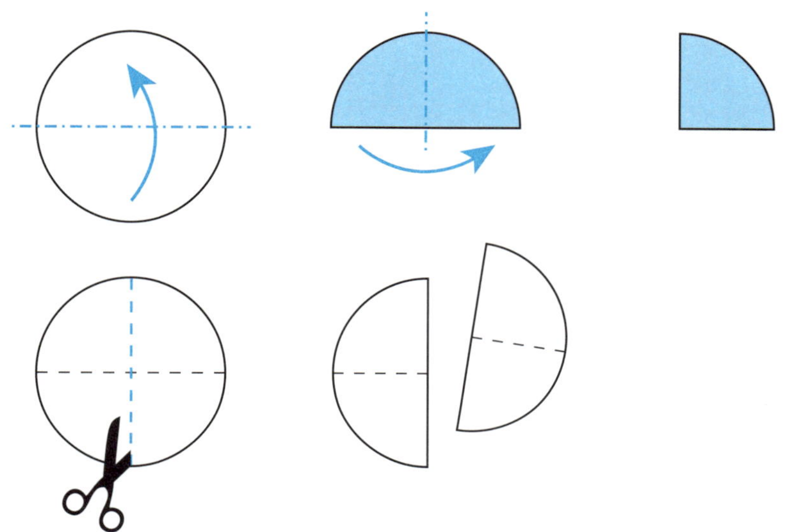

▲ *Fold and cut the circle to make semicircles.*

3. Say, *Look at one of the creases that shows a diameter. It divides the circle into two equal parts. Each part is called a semicircle. Every diameter divides a circle into two semicircles. Cut along a diameter to make two semicircles.* The students will use these semicircles in the next activity.

5. Quadrants and Sectors

Preparation

Complete Activity 4, above.

Activity

1. Review what the students know about semicircles. Say, *We can divide a semicircle into two equal parts as well. What type of line do we have to cut along?* (A radius.) Have the students identify the appropriate radius and direct them to make the cut. Say, *These two equal parts are called quadrants. If there are two quadrants in a semicircle, how many quadrants are in a circle?* (Four.) *Put one of your quadrants aside.*

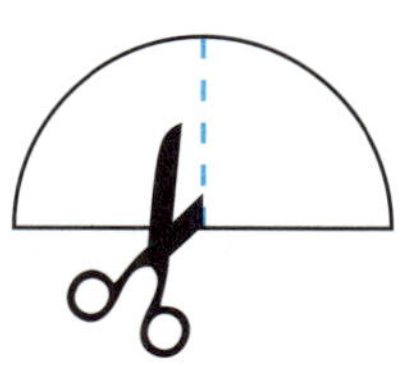

Materials

- Large sheet of paper and drawing compass — for demonstration
- Scissors — 1 pair for each student

Did You Know?

An arc is a part of any curve and is often used to mean part of a circle.

A diameter is a line segment that joins two points on a circle while passing through the center of the circle.

A radius is a line segment that joins the center of a circle with any point on its circumference.

A semicircle is a special arc of a circle, whose endpoints are the endpoints of a diameter. It is also thought of as the arc and the diameter together.

A quadrant is a special sector of a circle. It is formed by an arc and two radii that form a right angle. It covers one quarter of the area bounded by a circle.

A sector is the region bounded by two radii and an arc of a circle.

Materials

- The semicircles made by the students in Activity 4
- Drawing compass — 1 for each student
- Paper — 2 sheets for each student
- Scissors — 1 pair for each student
- Large sheet of paper — for demonstration
- Glue

Did You Know?

The words "sector", "section", and "secateurs" (another word for garden pruners or shears) share the same Latin root of *secare*, which means "to cut".

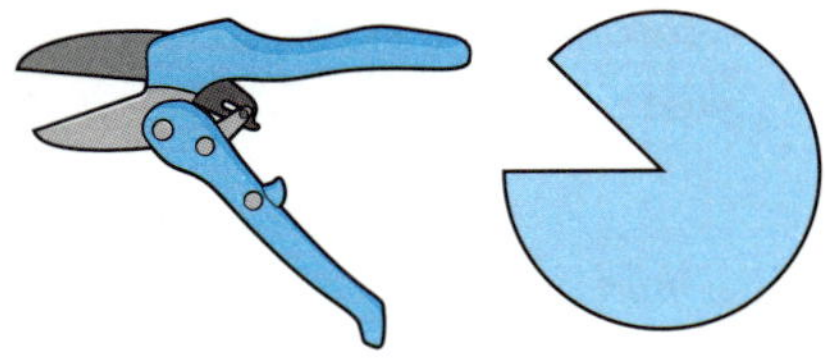

◼ Materials

- Paper, approx. 20 cm x 28 cm — 1 sheet for each student and 1 for demonstration
- Drawing pins — 2 for each student and 2 for demonstration
- Thick tagboard or heavy card board (the same size as the paper) — 1 sheet for each student and 1 for demonstration
- Piece of string or strong thread about 18 cm long — 1 for each student and 1 for demonstration

2. Say, *A circle can be divided into even smaller parts. The parts don't have to be the same size. Choose any point on the arc of your quadrant and draw a straight line from it to the center point of the circle. When you cut along this line you will create two sectors.* Bring out the ideas that semicircles and quadrants are both special types of sectors.

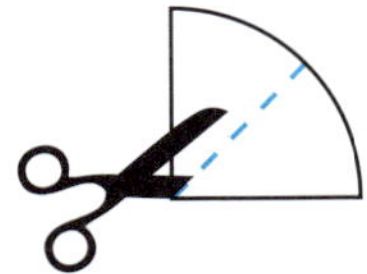 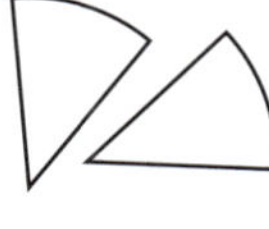

3. The students can label their circle parts and paste them onto a sheet of paper for reference.

6. Making Ellipses

Preparation

No preparation is required.

Activity

1. Direct each student to make a "hotdog" fold with his or her sheet of paper.

2. Have the students unfold the paper and draw two points about 5 cm apart along the crease, in about the middle of the sheet of paper. They can then stick a drawing pin through one of the points into the tagboard.

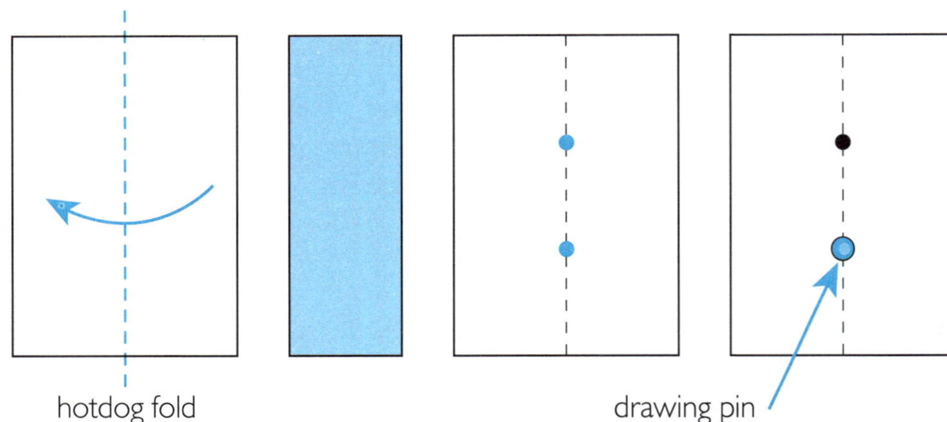

hotdog fold

drawing pin

3. Instruct the students to tie the ends of the string together and place the loop around the drawing pin. They then place the tip of a pencil inside the string loop, keep the pencil vertical, and push it gently against the string so that it is taut. By moving the pencil around the central point, keeping it inside the string, the students will draw a circle.

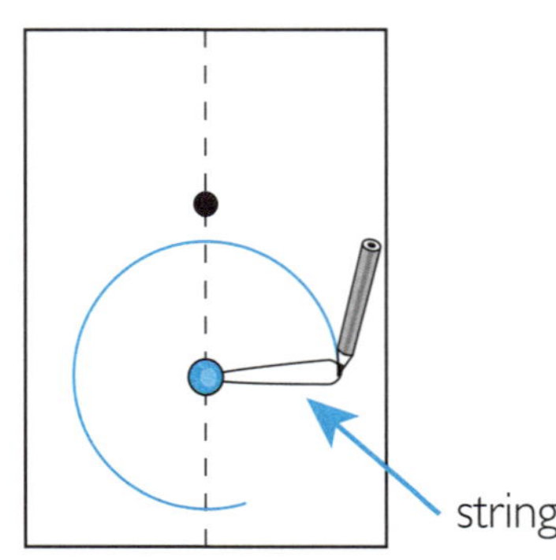

string

4. Review what the students know about circles. Bring out the fact that all points
   on a circumference are the same distance from the center. Say, *Identify the
   center of your circle. A diameter runs through the center of a circle.* Mark a diameter
   on your circle. *Can you have a diameter smaller than that?* (No, all the diameters
   of a circle are the same length.)

5. Direct each student to place his or her
   second drawing pin through the remaining
   point on their paper. Have the students
   predict and test what shape will be produced
   when they loop the string around both pins
   and move their pencils as they did when
   making the circle. Introduce the word
   "ellipse" to describe the resulting shape.

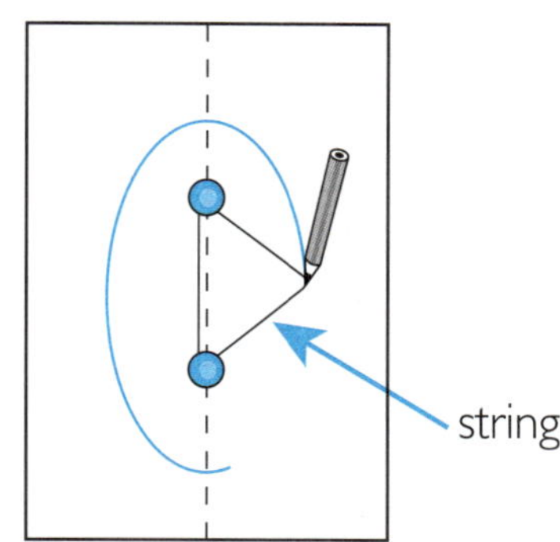

6. Encourage the students to predict the results of using longer and shorter
   pieces of string, and different numbers of drawing pins, and then experiment
   using those variations. The students will use these ellipses in the next activity.

7. Exploring Ellipses

Preparation
No preparation is required.

Activity

1. Discuss the ellipses that the students made in Activity 6. Hold up one example
   and ask, *What's the longest diameter you can draw on this ellipse?* (Draw a line
   through the two points where the drawing pins were to join opposite points on
   the ellipse.) *What's the shortest diameter?* (The shortest diameter is perpendicular
   to the longest one.) Introduce the term "focus" to describe the point where one
   of the drawing pins was. Introduce the plural of "focus" as "foci".

2. Ask, *What is the same and different about a circle and an ellipse?* (For example,
   a circle and an ellipse are both curves, but all points on an ellipse are not the
   same distance from a central point as is the case with a circle).

3. Direct the students to draw a straight line to join the foci. They can then join
   each of the foci to the same single point on the perimeter of the ellipse to form
   a triangle. Have the students measure the perimeter of this triangle. Have them
   record the answer and repeat with another point on the perimeter. Ask, *What
   do you notice?* (The totals are the same.) *Why do you think that occurs?* Bring
   out the idea that the loop of string used to make a given ellipse was a fixed
   length. The perimeter of any triangle made by joining the foci with a point
   on the perimeter of the ellipse will always be equal to this length.

Materials
- Ellipses from Activity 6

Did You Know?
The diameters in an ellipse are
called *axes*. There is a *major
axis* which is longest, and a
minor axis which is shortest.

When two lines intersect to
form a right angle they are
said to be perpendicular to
each other.

Materials

- Blackline Masters 7, 8, and 9 (pages 65–67) — 1 copy for each student
- Large compass — for the board
- Drawing compass — 1 for each student
- Color pencils — 1 set for each student
- Scissors — 1 pair for each student

8. Daisy Patterns

Preparation

No preparation is required.

Activity

1. Encourage the students to experiment with their compasses and create designs made by overlapping circles of the same size or different sizes. A useful way to start could be to have them examine what happens when the distance equal to the radius of a circle is progressively marked on the circle's circumference. Circles from these points can then be constructed. The first three designs below have been constructed using this method. You may like to demonstrate these on the board.

2. By using templates (such as pattern blocks) or drawing compasses to construct shapes (see Blackline Masters 7, 8, and 9), the students can use the vertices of a polygon as the centers of the circles.

3. The students may also like to shade the different regions in their pictures or make puzzle pieces by cutting along the lines of the design.

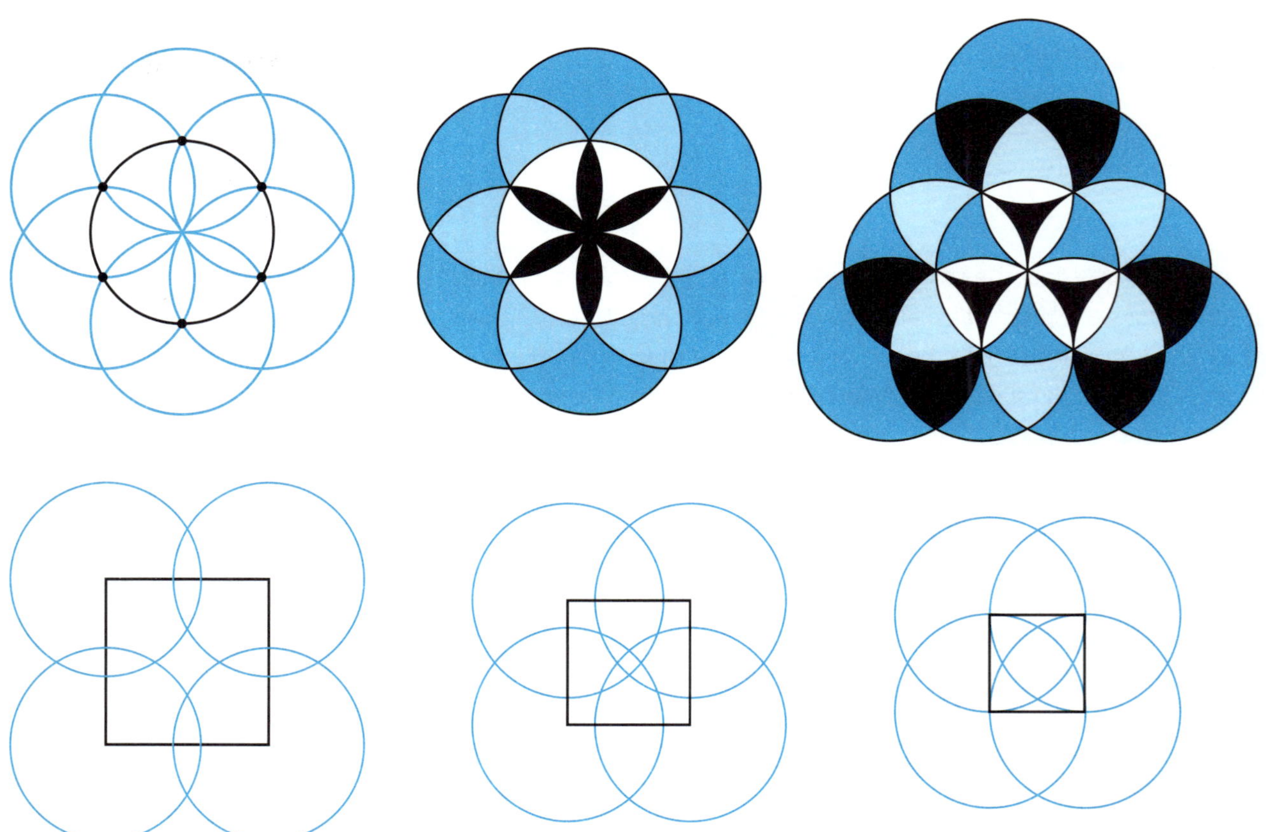

▲ *The students can overlap circles in particular ways to create many interesting designs.*

9. Concentric Circles

Preparation

No preparation is required.

Activity

1. Many dazzling designs can be made with concentric circles. Encourage the students to draw overlapping circles, arranging their centers on the vertices of other shapes, or varying the distance between each circle.

2. The students may also like to try combining concentric circles with the daisy patterns in Activity 8 ("Daisy Patterns").

Materials

- Drawing compass — 1 for each student

Did You Know?

The point where two sides of a polygon meet is called a "vertex".

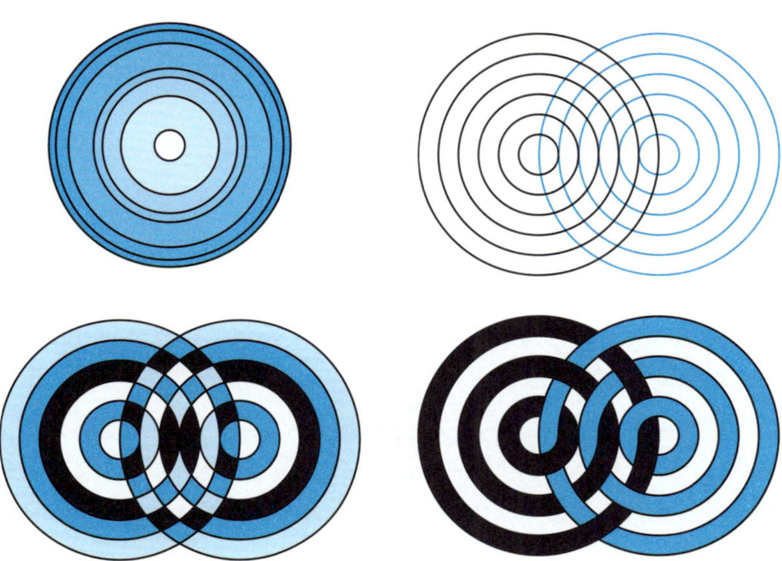

▲ *The students can make eye-catching designs using simple combinations of concentric circles.*

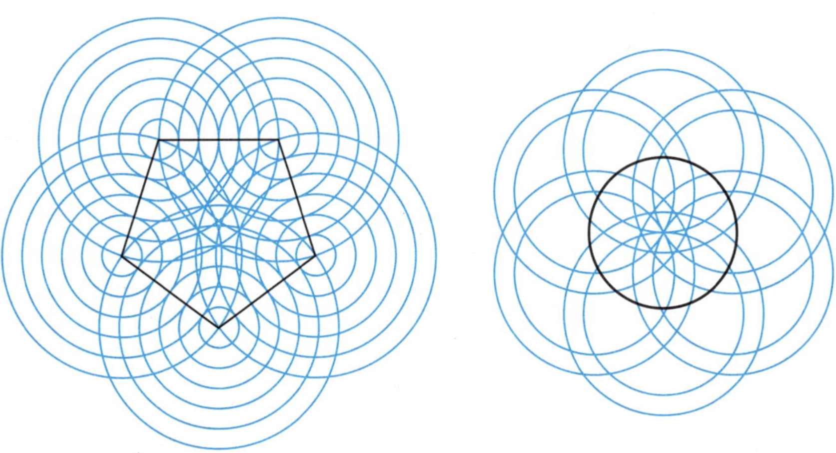

▲ *Many complex designs can be made by combining daisy patterns with concentric circles.*

triangles

Triangles are some of the simplest and most easily recognized polygons. Nonetheless, there are many interesting aspects to explore, such as those highlighted in the following activities.

■ Materials

- Blackline Masters 1 and 4 (pages 59 and 62)
- Scissors — 1 pair for each student
- Drinking straws — for each student
- Pipe cleaners — for each student

1. What is a Triangle?

Preparation

Copy Blackline Masters 1 and 4 onto tagboard or light card. Cut out the cards depicting triangles and discard the rest.

Activity

1. Display the cards to the students and ask, *What do these shapes have in common?* Some students may know that they are all called triangles, but you might also bring out the fact that the shapes all have three sides and three corners, and that the sides are straight. Emphasize that only closed shapes that have three straight sides can be called triangles.

2. Have the students use the straws and pipe cleaners to make three or four different triangles and two non-examples of triangles. Some are shown below. Discuss why the non-examples are not triangles (e.g. the shape does not have three sides).

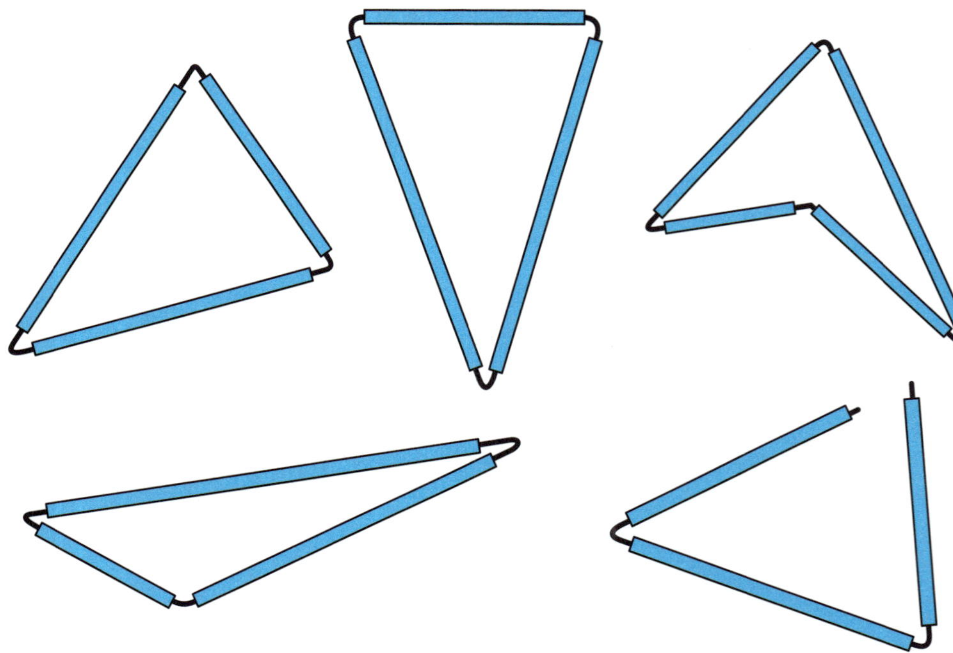

▲ *Have the students make examples and non-examples of triangles.*

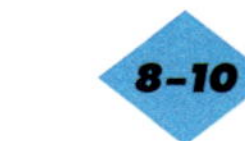 

Preparation

Distribute the materials to each pair of students. Have them cut the straws so that they have:

- 5 drinking straws that are 5 cm long
- 6 drinking straws that are 10 cm long
- 7 drinking straws that are 20 cm long

Activity

1. Challenge the students to investigate which combinations of straws will form triangles. They can use the pipe cleaners to join the sides of the triangles together. The possible solutions are shown below.

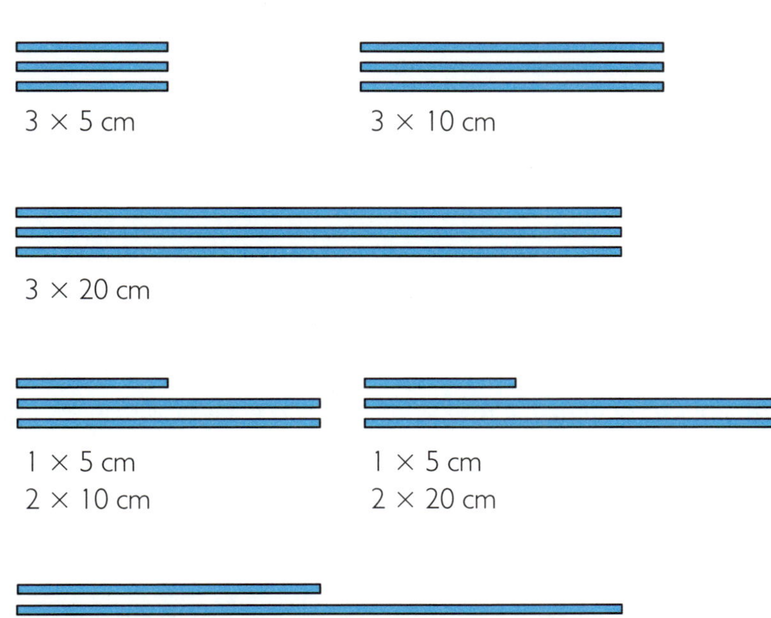

3 × 5 cm

3 × 10 cm

3 × 20 cm

1 × 5 cm
2 × 10 cm

1 × 5 cm
2 × 20 cm

1 × 10 cm
2 × 20 cm

▲ *Students will discover that these are the possible combinations
of straws that can be used to make triangles.*

2. After the students use models to discover all the solutions, have them figure out the solutions numerically. Ask questions such as, *Can a triangle be made using a 5-cm straw, a 10-cm straw, and a 20-cm straw? Why or why not?* The students can calculate the answer to this problem by figuring out whether the combined length of the two shortest straws (5 + 20) exceeds the length of the longest one (20). If the combined length is less than the longest side, a triangle cannot be made. Try other examples with your students, using sides of any length.

Materials

- Drinking straws, at least 20 cm long — 12 for each pair of students
- Pipe cleaners — for each pair of students
- Scissors — 1 pair for each pair of students

■ Materials

- Geostrips and fasteners or similar (e.g. drinking straws and pipe cleaners) for each pair of students

3. Types of Triangles

Preparation

No preparation is required.

Activity

1. Challenge each pair of students to make three different types of triangles using the geostrips: those with no sides the same length, those with two sides the same length, and those with three sides the same length.

2. Ask, *What names could you use to describe each type of triangle you made?* Discuss the students' ideas. Introduce the term "scalene" to describe triangles with sides that are unequal in length, "isosceles" for those with two equal sides, and "equilateral" for those with three.

3. Say, *Use your model as a template to draw a scalene triangle. Choose one angle on your drawing and compare it to the remaining two angle sizes. What do you notice?* (Each angle is a different size.) Ask, *Do you think all scalene triangles will have the same size angles? Compare your scalene triangle with one made by another pair of students.* The students will find that it will only be coincidence if the angles are the same.

4. The students can repeat the previous step with isosceles and equilateral triangles. They will find that there are two identical angle sizes in any isosceles triangle, but that any two isosceles triangles will only coincidentally have the same angle sizes. On the other hand, the students will discover that not only does any equilateral triangle have all angles of the same magnitude, any two equilateral triangles will have exactly the same angle measurement for each interior angle, that is, 60°.

Did You Know?

"Isosceles" comes from the Greek words *isos*, meaning "equal", and *skelos*, which means "leg". The word "scalene" comes from the Greek word *skalenos*, which means "uneven". "Equilateral" has its roots in the Latin words *æquus*, which means "equal", and *latus*, which means "side".

geo For other interesting activities on different types of triangles see *Paper Polygons* and *All About Angles*.

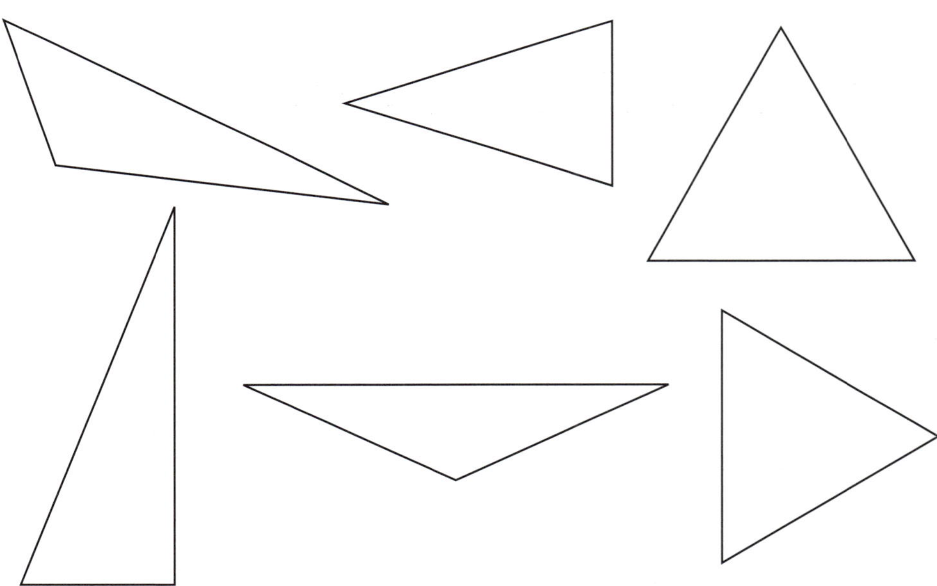

▲ *Have the students investigate the angles in scalene, isosceles, and equilateral triangles.*

4. Angles in a Triangle

Preparation
No preparation is required.

Activity
1. The students should be comfortable with using protractors before completeing this activity. Demonstrate the following method for drawing triangles using a protractor.

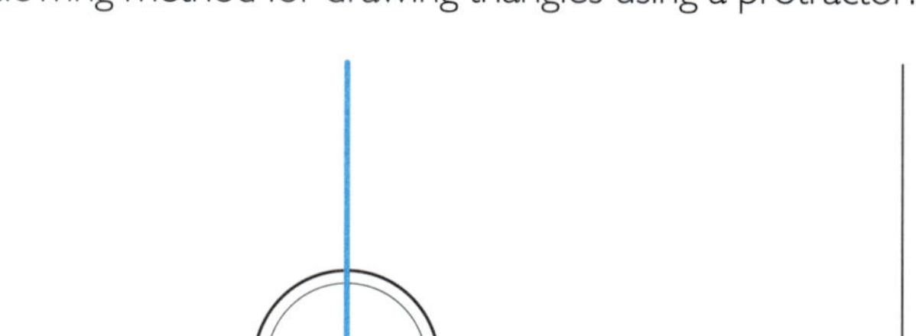

a. Draw a line. This is the first side of the triangle. Mark a vertex (point A).

b. Align the protractor on point A. Mark 90° and draw a line to form the second side of the triangle.

c. Mark a second vertex (point B) on the original line.

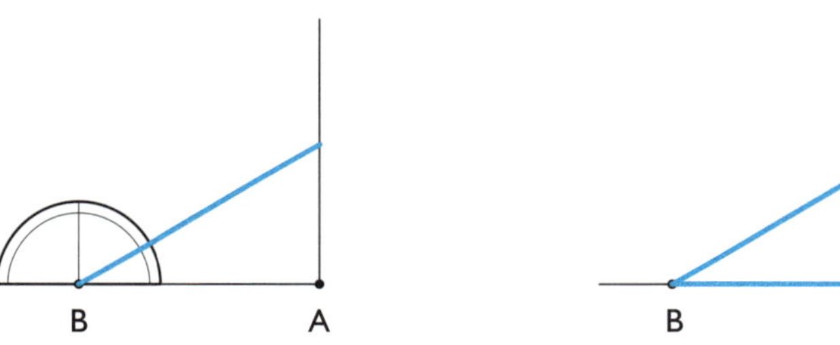

d. Align the protractor on point B. Mark 30° and draw a line to form the third side of the triangle.

▲ *Demonstrate these instructions for constructing a triangle with sides of different lengths. Shown is a 30°, 60°, and 90° scalene triangle.*

2. Direct each student to construct triangles that have the following angles. Have them label each angle in each triangle. The size of the triangles is unimportant; do not dictate the lengths of sides. The students will use the triangles they have constructed in the next activity, "Similar Triangles".

- 30°, 60°, 90°
- 50°, 60°, 70°
- 20°, 50°, 110°
- 40°, 40°, 100°

3. As the students construct the triangles, ask, *What do you notice after you measure two angles? Do you need to measure the third angle?* (No.) *Why do you think that happens?* Discuss how the sum of each set of angles written on the board is 180°. As there are only three angles inside a triangle, when the sides that make two angles are drawn, the third angle is automatically made.

Materials
- Protractors — 1 for each student
- Large protractor — for the board

Did You Know?
The word "protractor" comes from two Latin words — *pro*, meaning "forward", and *trahere*, meaning *"draw or drag"*. To use a protractor, it has to be imagined that a line is being dragged or turned from one position towards another.

■ Materials

- The triangles the students drew in Activity 4
- Scissors — 1 pair for each student
- Blu-Tack

5. Similar Triangles

Preparation

No preparation is required.

Activity

1. Have the students carefully cut out the triangles that they made in Activity 4 and write their names on their triangles.

2. Arrange the students into small groups and have them sort their triangles into collections according to the size of the angles. Then follow the steps below for each collection of triangles.

    a. Identify the largest triangle.

    b. Identify the corner formed by the shortest and longest sides.

    c. Identify this same corner in the next-largest triangle. Place this triangle on top of the largest so that their corners align perfectly. Use Blu-Tack to keep the triangles in position.

    d. Repeat the previous step with the next-largest triangle and so on.

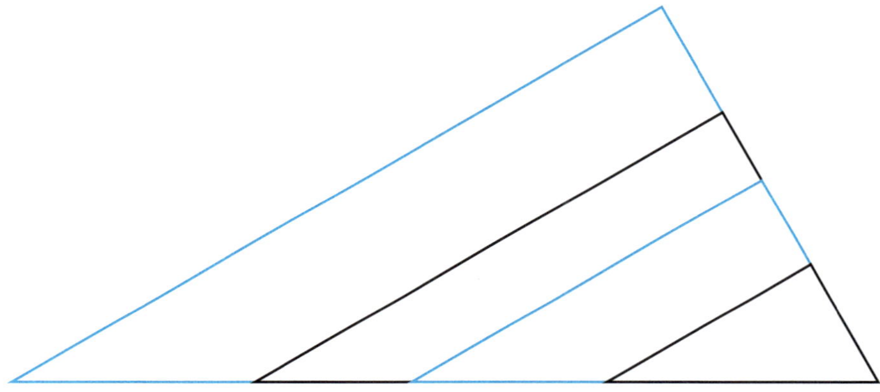

▲ *The students should place the triangles so that one particular corner in each is perfectly aligned to create a stack.*

3. Invite the groups to show their triangle stacks. Call upon students to describe what they notice about each stack of triangles. Some students may point out that the second-longest side of each triangle in a stack is parallel with all the others in that stack. Others may simply say that all the triangles in a stack look the same, despite being different sizes. This concept can be highlighted by placing triangles with different angles on top of one another. Encourage the students to think of a rule they could use to describe what they noticed. For example, *All triangles with the same angles have the same shape.*

Did You Know?

"Similar" is a word used to describe shapes that are exactly the same shape but are different sizes. In mathematics, it's used more precisely than in everyday language, where it has a slightly different meaning.

geo For many more investigations into angles, see *All About Angles*.

quadrilaterals

There is a greater variety of names for particular four-sided shapes than for any other two-dimensional shape. The discussion on page v of the Introduction to this book highlights some potential variety in terms. As you work through the following activities, be sure to use the correct terminology for the region in which you teach.

1. What is a Square?

Preparation
No preparation is required.

Activity

1. Hold up four uncut straws and gain the students' agreement that they are all the same length. Ask, *What shapes can we make by joining these straws together?* Discuss the students' answers, before joining the straws together with pipe cleaners to make a square. Ask, *What is the name of this shape?* (Square.) *What is special about the length of each side?* (They are all the same length.)

2. Say, *There is something else that is special about a square.* Make another square and push the "top" to one side so that it no longer looks like a square. Ask, *Does it look like a square now?* (No.) *Are the sides all the same length?* (Yes.) *So why is it not a square any more?* (The corners have different shapes.) Retain the square you have constructed for use in the next activity, "What is an Oblong?".

3. Have the students compare the square pattern blocks and elicit their agreement that each of the four corners are the same. Position the straw square where the students can see it and ask, *This shape looks like a square, but how can we use the square pattern blocks to test whether it really is a square?* Discuss the students' suggestions and bring out the idea that the pattern blocks can be placed against the inside of each corner of the square straw. If there is no gap between the blocks and the straws, then the corners are the same. Place the pattern blocks against the square you changed to demonstrate this point.

4. Encourage volunteers to describe how to test if a shape is a square. An answer (with guidance) may be, *If there are four straight sides, and all sides are the same length, and all the corners are the same, then the shape is a square.*

5. Have the students use the straws and pipe cleaners to make two squares and two shapes that are "almost" squares. Encourage the students to use the pattern blocks to test each of the corners. Ask individual students to explain why their shapes are or are not squares.

Materials
- Drinking straws
- Pipe cleaners
- Square pattern blocks

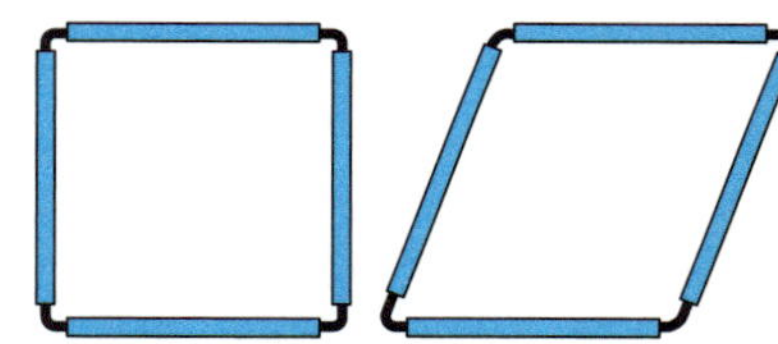

▲ *Push the "top" of the square to show that a shape must have four equal sides and four equal corners to be called a square.*

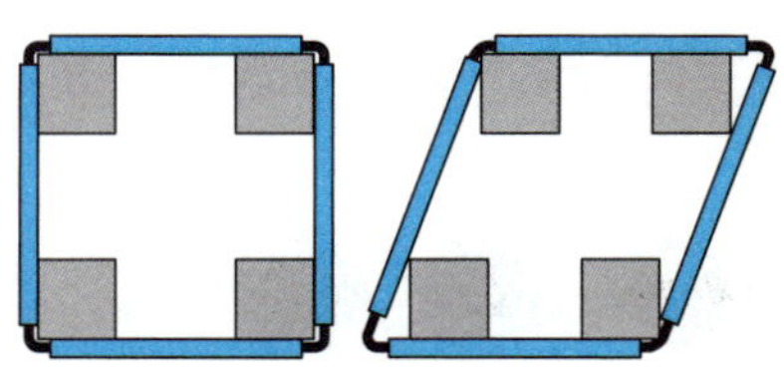

▲ *Place square pattern blocks inside the straw shapes to show how to test if the corners in each shape are equal.*

Materials

- The drinking-straw square made in Activity 1
- Extra drinking straws
- Scissors — 1 pair for each student
- Square pattern blocks
- Pipe cleaners

Did You Know?

An oblong is a rectangle with adjacent sides of different lengths. It is also known as a "non-square rectangle".

2. What is an Oblong?

Preparation

No preparation is required.

Activity

1. This activity follows on from the previous one. Review what the students know about squares. Cut a straw in half and discuss with the students how the two sections are the same length. Hold them beside another two whole straws and compare their lengths.

2. Ask the students to predict what shapes can be made using the straws (an oblong and kite). Make an oblong and ask, *What can you tell me about this shape?* Students may say that it is a rectangle, that it has four straight sides, and that the corners look to be the same shape. Ask, *How can we check that the corners are the same shape?* Students may suggest using pattern blocks, as described in Activity 1.

3. Hold up the oblong and the straw square from Activity 1 and ask, *What is the same about these two shapes?* (They both have four sides, the sides are straight and all of the corners are the same.) *What is different about them?* The students will most likely suggest that the oblong has "two short sides and two long sides". Say, *Another way to say this is that opposite sides are the same length.*

4. Draw a comparison between types of cars and types of rectangles. Say, *A sedan and a limousine are both types of cars, but one is longer than the other. In the same way, both of these shapes are types of rectangles, but one of them is longer than the other. The longer one is called an oblong. The other one is called a square.*

5. Invite the students to use the materials to make oblongs and non-oblongs of different sizes. Ask volunteers to explain why their shapes are or are not oblongs.

▲ *Both of these shapes are types of rectangles. The one on the left is an oblong and the one on the right is a square.*

Materials

- Blackline Master 6 (page 64)
- Geostrips and fasteners, or geoboards and rubber bands — 1 set for each student

3. What are these Quadrilaterals?

Preparation

No preparation is required.

Activity

1. Some experience with parallel lines is a prerequisite for this activity. Introduce the term "quadrilateral" to describe polygons with four sides. Say, *Make some different quadrilaterals that you can name.*

2. Discuss the shapes that the students have made: they should include oblongs and squares. Quadrilaterals that the students may have seen before but can not name could include kites, rhombuses, and trapezoids. Use the following guidelines to assist the students in making and naming them.

Kite

Say, *Make an oblong. Now, make a quadrilateral that is not an oblong but has two sides of one length and two sides of another length.* The students will produce a "kite" or a non-rectangular parallelogram. Find and hold up an example of a kite and introduce the name.

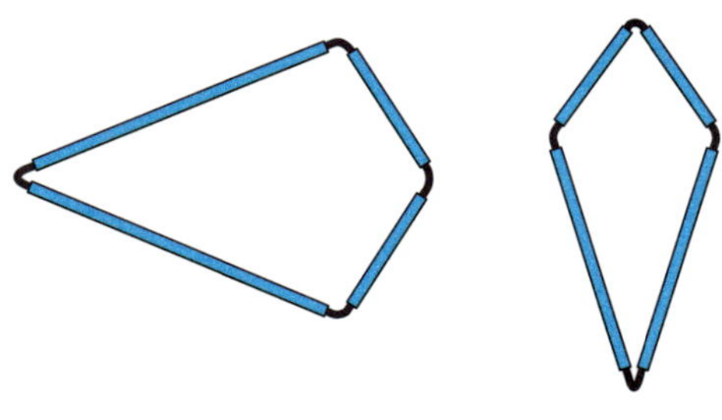

Rhombus

Say, *Make any quadrilateral that has four equal sides.* Encourage the students to make a shape that is not a square, as that is what many will do initially. Explain that any quadrilateral that has four equal sides is called a "rhombus". A square is a special type of rhombus in the same way as a Chihuahua is a special type of dog.

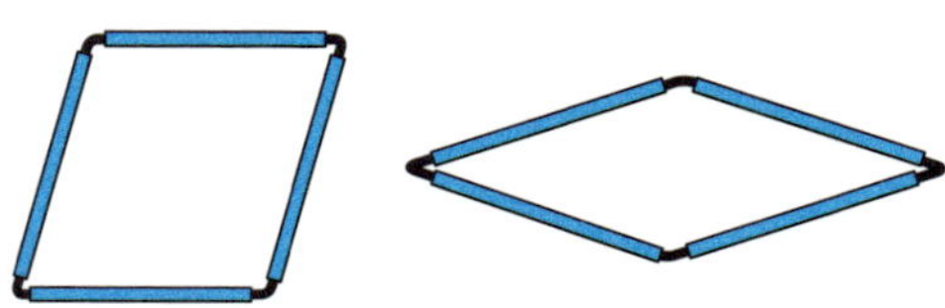

Parallelogram

Like the rhombus, this shape provides an opportunity for discussion. Say, *Make any quadrilateral where opposite sides are parallel.* The students may create squares, oblongs, or rhombuses. All of these shapes fall into the category of "parallelograms". Some students may make the shape that is usually associated with the term "parallelogram", as shown at right.

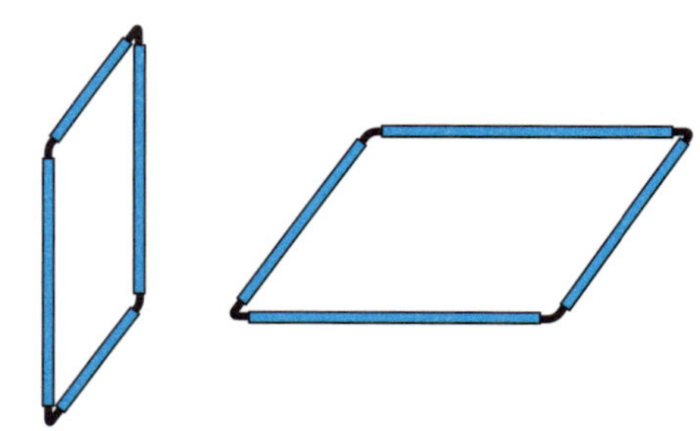

Trapezoid

Say, *Make a quadrilateral where only two sides are parallel.* The students will produce a shape that is a trapezoid. Hold up an example and introduce the name (see the note in the margin).

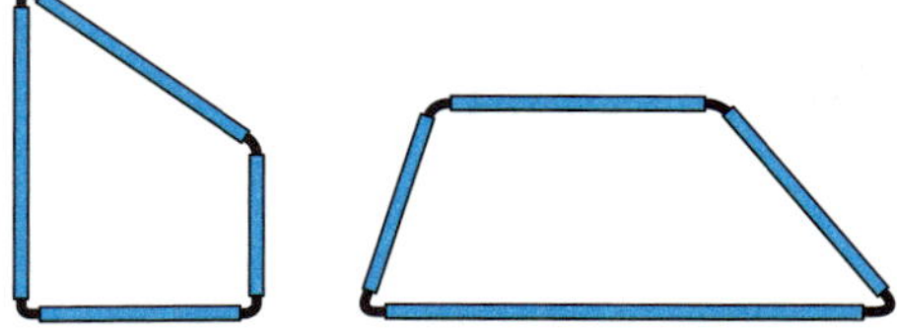

3. The students may be interested to learn about the roots of the shapes' names, so share some of the information from Blackline Master 6 (page 64) with them.

See *All About Angles* for more activities involving parallel lines.

Materials

- Geostrips and fasteners, or geoboards and rubber bands — 1 set for each pair of students

4. Quadrilateral Comparisons

Preparation

No preparation is required.

Activity

1. Direct the students to work in pairs to make examples of a square, oblong, kite, rhombus, and trapezoid.

2. Position examples together so that the students can see them. Compare the kite to an oblong. Ask, *What is the same about a kite and an oblong?* (They have four straight sides, they both have two sides of one length and two of another length, etc.) *What are the differences between a kite and an oblong?* (An oblong has opposite sides the same length and four right angles, a kite has adjacent sides the same length and may or may not have right angles.)

3. Repeat Step 2 for a kite and a square, then a rhombus and a trapezoid. Then ask, *What makes a kite a kite?* The students may need prompting to remind them of what they have observed. A simple explanation is that a kite is a quadrilateral with two pairs of equal adjacent sides and one pair of opposite angles that are equal.

4. Direct the students to compare and contrast all of the shapes and to record their observations. As they compare the shapes, direct them to write descriptions or definitions for each one. A table like the one below can help them keep a record of which shapes they have compared.

	Oblong	Square	Kite	Trapezoid
Rhombus	✓	✓		
Trapezoid			
Kite		
Square	

▲ *Using a table like this can help the students keep a record of which shapes have been compared.*

Materials

None

5. Midpoints

Preparation

No preparation is required.

Activity

1. Ask the students to draw any quadrilateral; the shape is unimportant. They can then measure to locate the midpoint of each side.

2. Ask, *What shape will you see when you join the midpoints on your quadrilateral?* Have each student join the midpoints on his or her shape and compare the result with a partner's shape.

3. Discuss the results. The students should report that they have some type of parallelogram, whether it is a square, oblong, rhombus, or otherwise. Highlight the fact that all of these shapes fit the definition of a parallelogram, that is, a quadrilateral with opposite sides equal and parallel.

4. Invite the students to use a range of other quadrilaterals. Encourage them to create shapes that are concave, that have great differences in side length, or that are more familiar, such as squares and kites. They will find that the result is always the same.

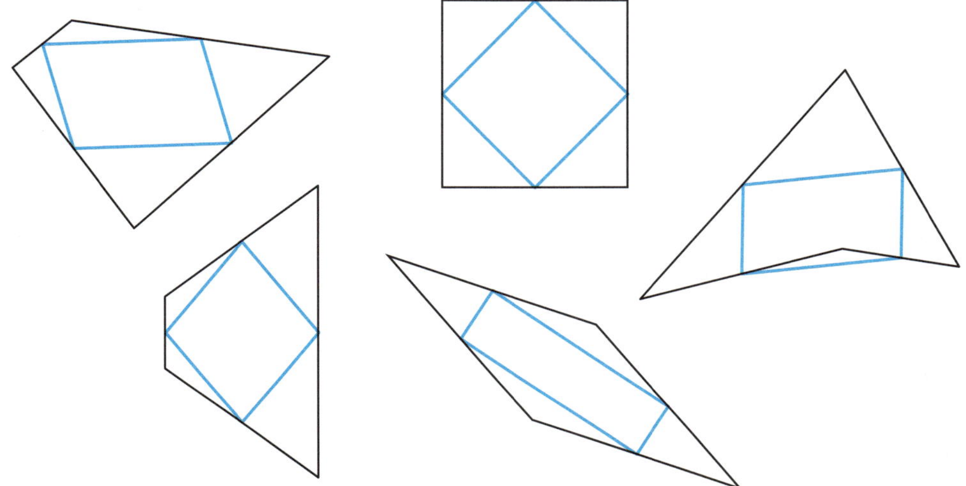

▲ *The students will discover that joining the midpoints on the sides of a quadrilateral will always form a type of parallelogram.*

6. Angles in a Quadrilateral

Preparation
No preparation is required.

■ Materials
None

Activity
1. Direct pairs of students to draw six convex quadrilaterals each. They can then count how many interior angles in each shape are either equal to, greater than, or less than a right angle.

2. Instruct each pair of students to determine the combinations of angles that occur. Discuss the results as a class and bring out the following points:
- *If their shape has only two angles greater than a right angle, then the remaining angles are less than a right angle.*
- *If their shape has only two right angles, then one of remaining angles is more than a right angle and one is less.*
- *If their shape has three right angles, then the fourth must also be a right angle.*

Although there are many different types of pentagon, this unit focuses on regular pentagons. A regular pentagon has all angles equal in magnitude and all sides equal in length. The special properties of this particular shape provide opportunities for numerous fascinating explorations.

■ Materials

- Blackline Master 20 (page 78)
- 2 blank transparency sheets
- Tagboard or light card —
  1 sheet for each student
- Overhead projector
- Overhead transparency pen

Did You Know?

Composite shapes are created by joining together other, smaller shapes.

Concave polygons have at least one interior angle that is larger than 180°; that is, they have a reflex angle.

A diagonal line is any straight line segment joining two non-adjacent vertices in a polygon.

An isosceles trapezoid is one in which the non-parallel sides are the same length.

An isosceles triangle has two sides of equal length and two equal angles. The angles opposite equal sides are equal in measure.

1. Shapes Within Shapes

Preparation

1. Make an overhead transparency (OHT) of Blackline Master 20.

2. Make a copy of Blackline Master 20 on a sheet of tagboard or light card for each student.

3. Cut the student copies of Blackline Master 20 in half so that each student has one pentagon. Retain second pentagon on the blackline master for the next activity, "Creating Pentagon Pictures".

Activity

1. Define a diagonal line as one which joins two vertices that are not next to one another (i.e. they are non-adjacent). Display the OHT and instruct each student to draw diagonal lines to connect all the vertices inside the pentagon while you do the same on the OHT.

2. Discuss the shapes that the students can see inside the pentagon. Many students may see a star. Some students may identify another regular pentagon and some triangles. Older students may identify the triangles as isosceles triangles.

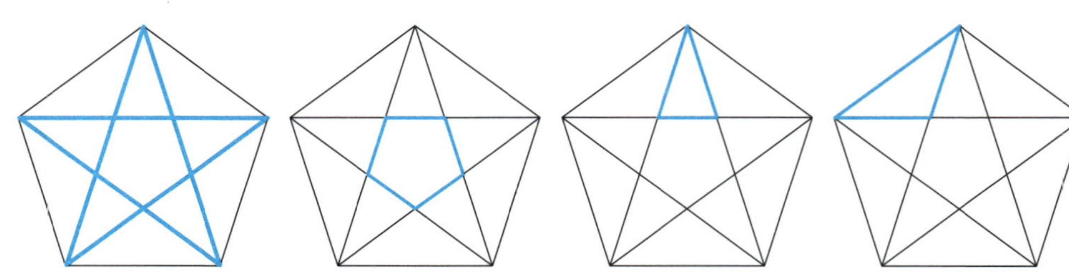

▲ *Discuss the shapes the students can see inside the pentagon.*

3. Encourage the students to consider what shapes can be seen if two or more of these internal shapes are joined together. For example, larger versions of the two different isosceles triangles can be seen, along with an isosceles trapezoid and a concave pentagon. Challenge the students to find as many different shapes as they can and sketch them.

4. Afterward, invite volunteers to show their results to the class. Have them place the spare blank overhead transparency over the photocopied one and trace the outline of each shape they find.

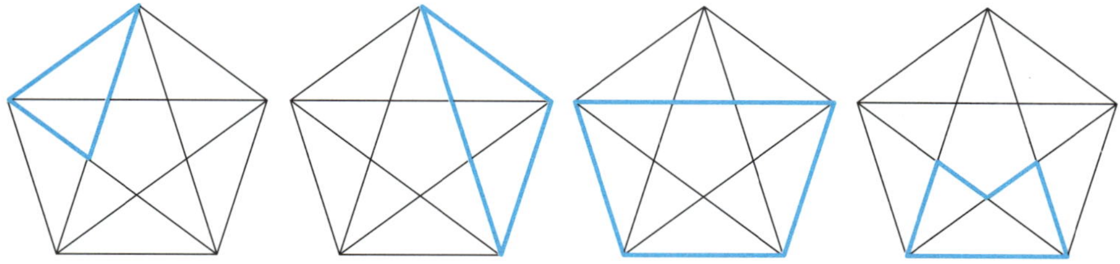

▲ *Challenge the students to find composite shapes.*

2. Creating Pentagon Pictures

Preparation
Make copies of Blackline Master 20 on tagboard or light card and cut in half so that each student has a pentagon (or use the copies left over from Activity 1, "Shapes Within Shapes").

Activity
1. Direct each student to draw diagonal lines to connect the vertices of his or her own polygon. The students can then cut out each smaller shape that results. They should each have a smaller pentagon and ten triangles.

2. Encourage the students to rearrange the shapes to form other shapes and pictures. Whether they use only a few shapes or all of them, discuss the results and have the students display their designs.

■ Materials
- Blackline Master 20 (page 78)
- Scissors — 1 pair for each student
- Sheet of paper — 1 for each student
- Glue — for each student

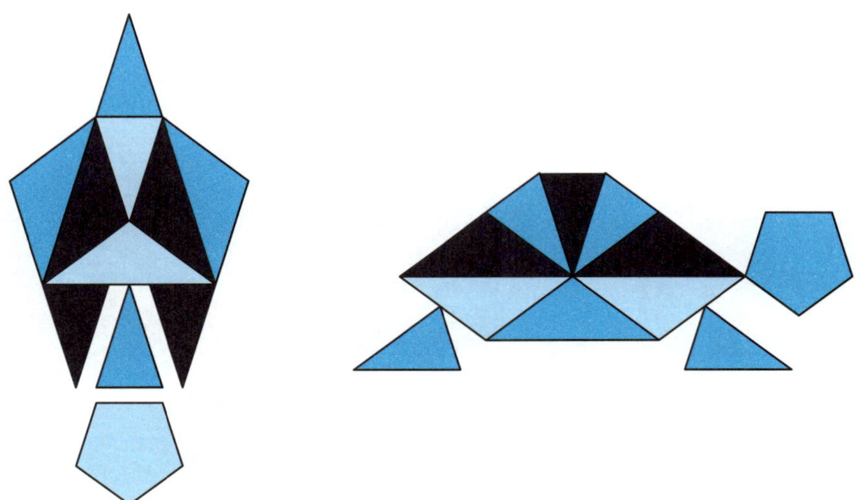

▲ *The students will enjoy making designs using the shapes found within a pentagon.*

■ Materials

- Blackline Master 20 (page 78)
- Tagboard or light card —
  2 sheets for each student
- Scissors — 1 pair for
  each student
- Large envelope, plastic zip-lock
  bag, or similar — 1 for each
  student
- Scrap paper — 2 or 3 sheets
  for each student

Did You Know?

A diagonal line is any straight line segment joining two non-adjacent vertices in a polygon.

An isosceles triangle has two sides of equal length and two equal angles. The angles opposite equal sides are equal in measure.

An acute triangle is one in which the largest angle is acute (less than 90°).

An obtuse triangle is one in which the largest angle is obtuse (more than 90° but less than 180°).

3. Dazzling Designs

Preparation

Make a copy of Blackline Master 20 on a sheet of tagboard or light card for each student.

Activity

1. Direct the students to cut out one of the two polygons on the handout. On one pentagon, have them draw diagonal lines to connect the vertices. They can then cut out each smaller shape that results: a smaller pentagon and ten triangles. All of these pieces and the other larger pentagon will be used for other activities. Tell the students to store them in an envelope or similar.

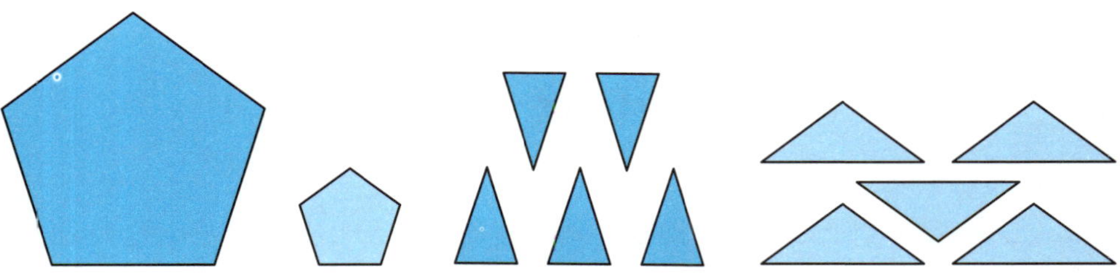

▲ *After cutting up one pentagon, the students will have 12 shapes in total.*

2. Have each student draw around the larger pentagon onto the scrap paper. The students can then investigate how multiple copies of each type of smaller shape can be arranged inside the large pentagon without overlap. That is, the students can examine how the small pentagon can be arranged, how the obtuse isosceles triangle can be arranged, and how the acute isosceles triangle can be arranged. The diagram below shows some possibilities. The students will find that that the acute isosceles triangle is the most versatile.

3. As the students investigate the shapes, they can position the smaller shapes together and draw around the edge of the resulting shape. After a number of designs are made, have the students select one and copy it onto the tagboard or light card for display.

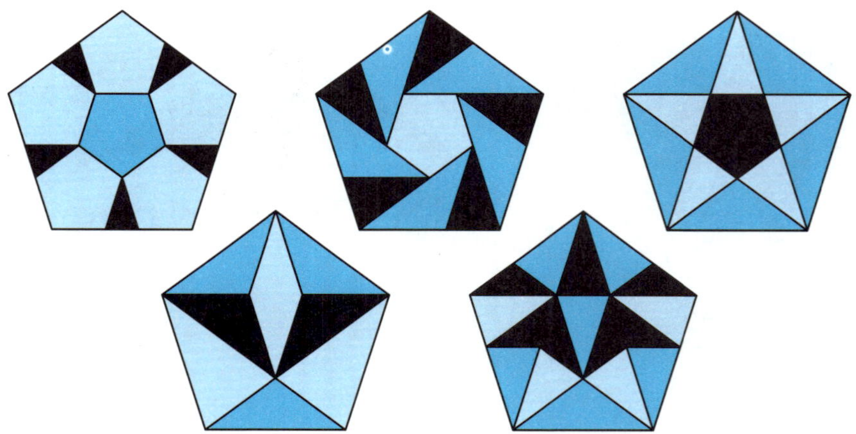

▲ *The students will be intrigued by the fascinating designs that are possible by repeatedly copying just one shape.*

4. Leaving No Gaps

Preparation
No preparation is required.

Activity

1. Talk about all of the pentagon pieces that the students have. (Each student should have a large pentagon, a small pentagon, and ten triangles.) Challenge the students to experiment with the different ways that they can cover a surface by tracing around the smaller pentagon on the scrap paper. Ask, *What do you notice?* Bring out the fact that unless the outlines overlap, there is no way of tiling the pentagons without leaving gaps.

2. Direct the students to put their pentagons to one side. The students can work in pairs or small groups to try using the triangles to cover a surface. Some may create a random arrangement of shapes, while others may make a repeating pattern, or tessellation.

3. After a few trials, have the students recreate one design on the tagboard or light card for display.

■ Materials
* Students' pentagon pieces from Activity 3
* Scrap paper — 1 sheet for each student
* Tagboard or light card — 1 sheet for each pair or group of students

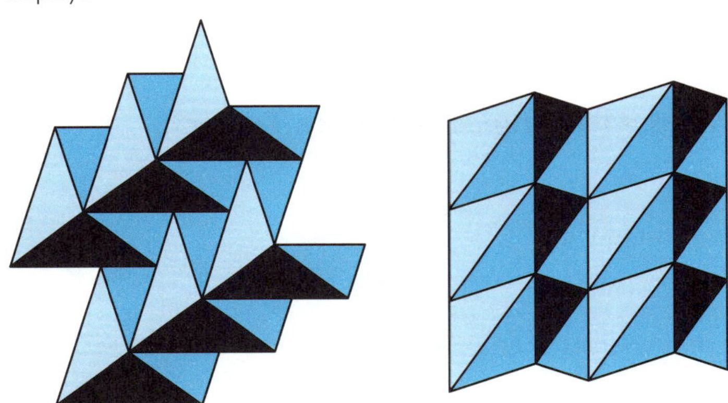

▲ *Although the regular pentagon will not tessellate by itself, the triangles will.*

5. Holey Pentagons

Preparation

1. Copy Blackline Master 20 (page 78) onto tagboard or light card.

2. Inside one of the pentagons, draw diagonal lines to connect the vertices and cut out the resulting shapes.

Activity

1. With the overhead projector turned off, arrange the acute isosceles triangles to make a complete pentagon, as shown in the diagram below.

■ Materials
* Blackline Master 20 (page 78)
* Sheet of tagboard or light card
* Scissors — for demonstration
* Overhead projector
* Students' pentagon pieces from Activity 3
* Blu-Tack

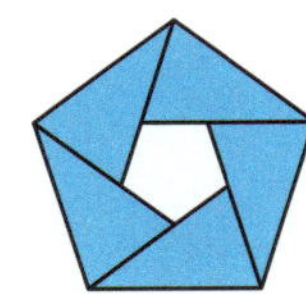

▲ *A "holey pentagon" made from acute isosceles triangles.*

2. Turn on the projector and set the students the following challenges:

- *Use only the acute isosceles triangles to make a pentagon that has another pentagon inside it.*
- *Use only the obtuse isosceles triangles to make a pentagon that has another pentagon inside it.*

3. The students may find it easier if they use little bits of Blu-Tack to keep the shapes in position. If the students need a hint for the first problem, remove one of the triangles from the overhead projector. If they have trouble with the second challenge, suggest that they try overlapping the pieces. If no students are able to accomplish the tasks, use your pentagon pieces on the overhead projector to reveal the solutions and have the students repeat your actions with their own pieces.

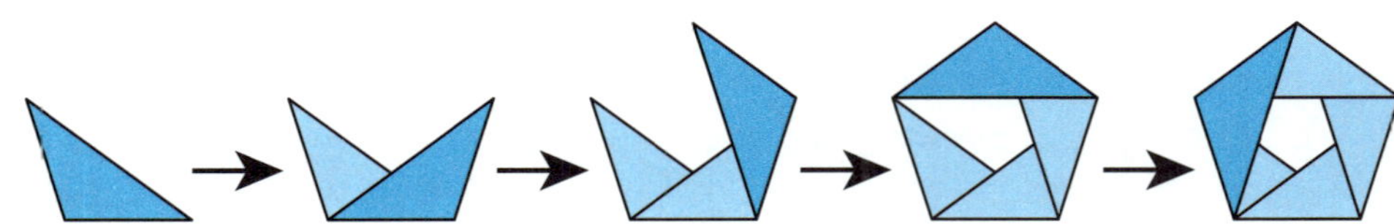

▲ *A "holey pentagon" made from overlapping obtuse isosceles triangles.*

■ Materials

- Students' pentagon pieces from Activity 3

Did You Know?

An acute triangle is one in which the largest angle is acute (less than 90°).

An obtuse triangle is one in which the largest angle is obtuse (more than 90° but less than 180°).

Congruent angles have the same measure, regardless of the length of their angle arms.

6. Working with Angles — Acute Triangle

Preparation

No preparation is required.

Activity

1. Discuss the properties of isosceles triangles. Highlight the fact that all isosceles triangles have two equal sides and two equal angles.

2 Challenge the students to figure out the size of the smaller angle in the acute isosceles triangle without using a protractor, but by arranging the triangles in particular ways.

3. If necessary, discuss how one full turn around a point is equal to 360°. Encourage the students to apply this fact by arranging copies of the triangle around a point, as shown in Figure A below. Such an arrangement shows that five of the angles together are equal to half a full turn; that is, they are equal to 180°. Therefore, each angle must be one-fifth of 180, so each of the smaller angles is equal to 36°.

4. Invite the students to calculate the larger angles in the acute triangle. Figure B shows a method similar to that explained earlier. If the five congruent angles cover 360°, then each must be 72° (360 divided by 5 equals 72). Alternatively, two of the smaller angles can be superimposed on the larger angle exactly, therefore the larger angle is 72° (36 multiplied by 2 equals 72).

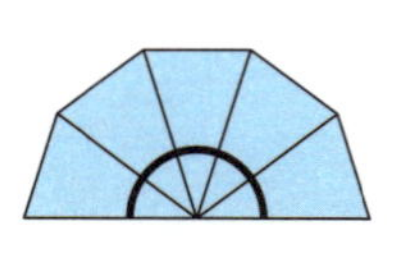

Figure A

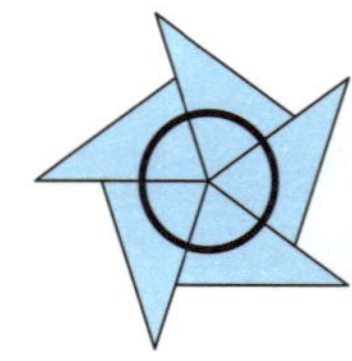

Figure B

▲ *These arrangements can be used to calculate the size of the angles in this acute triangle.*

7. Working with Angles — Obtuse Triangle

Preparation

No preparation is required.

Activity

1. As a follow-on from the previous activity, encourage the students to work out the angles in the obtuse triangle. The simplest method is to overlap the known angles of the acute triangle with the unknown angles in the obtuse triangle.

2. A more challenging approach is to arrange the copies of a particular angle around a point. In the diagrams below, Figure A shows an arrangement that can assist in figuring out the smaller angle. Since all the angles together form 180°, each must be equal to one-fifth of 180, which is 36°. Using what they know about the angles of the acute isosceles triangle, the students may use an arrangement like Figure B, and calculate that three of the larger angles in the obtuse angle together must be equal to 360 less 36. Therefore, each of the larger angles is equal to 108° (360 less 36 is 324, divided by 3 is 108).

Materials
- Students' pentagon pieces from Activity 3

More activities that involve calculating the magnitude of angles without using a protractor can be found in *Paper Polygons* and *All About Angles*.

Figure A   Figure B

▲ *These arrangements can be used to calculate the size of the angles in this obtuse triangle.*

8. Working with Angles — Pentagon

Preparation

No preparation is required.

Materials
- Students' pentagon pieces from Activity 3

Activity

Leading on from the previous two activities, the students may be interested to calculate the interior angles of the original regular pentagon they cut up. One way to figure this out is to recognize that a corner of the pentagon is made up of the smallest angles of two obtuse triangles and one acute triangle. Adding these three angles together gives a total of 108° (36 + 36 + 36 = 108). Alternatively, the students may notice that any of the corners of the pentagon can be overlayed by the larger angle of the obtuse triangle and so equal 108°.

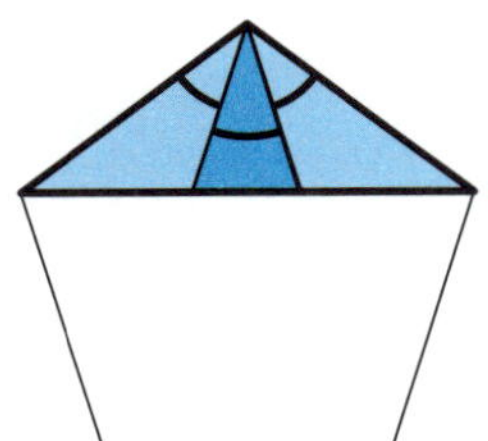 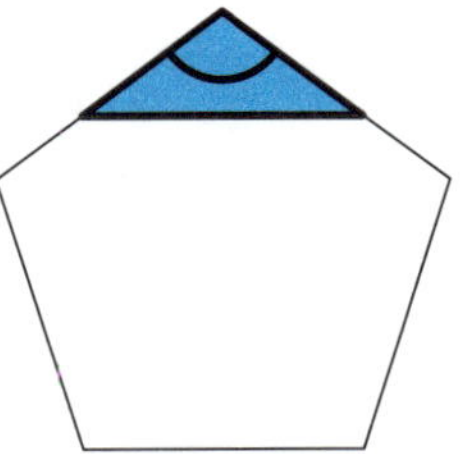

▲ *Using the known angles of the isosceles triangles, the students can calculate the interior angles of a regular pentagon.*

■ Materials

- Blackline Master 20 (page 78)
- Overhead projector and blank transparency sheet
- Students' pentagon pieces from Activity 3
- Calculator — 1 for each student

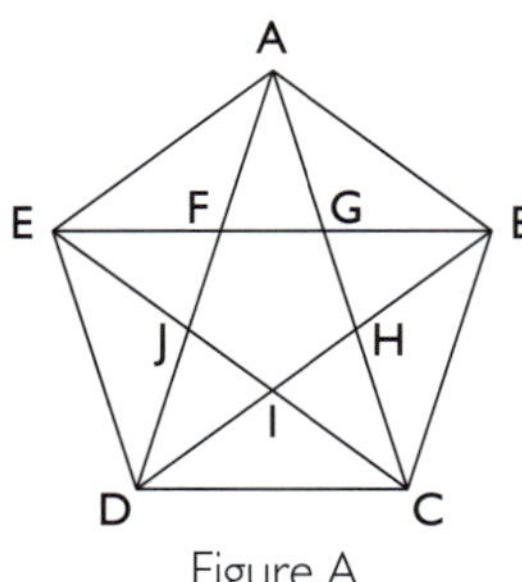

Figure A

▲ *Have the students label these points to help describe the shapes they see.*

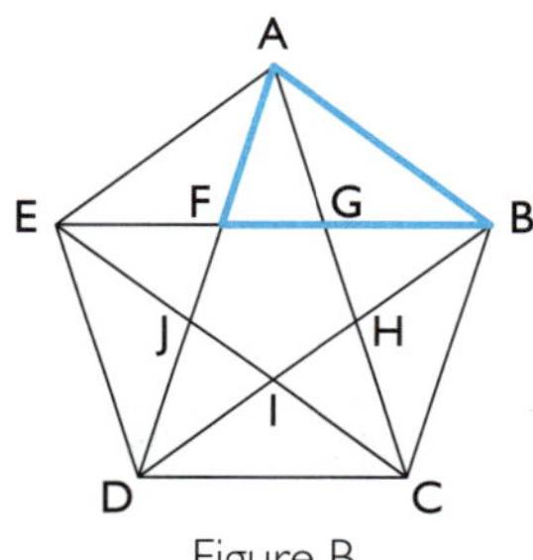

Figure B

▲ *The triangle shown can be described as Triangle ABF.*

Did You Know?

A "vertex" is the point where two or more line segments join or intersect on a 2D shape.

"Similar" is a word used to describe shapes that are the same but different sizes. In mathematics, it's used more precisely than in everyday language, where it has a slightly different meaning.

9. Finding Similar Shapes

Preparation

Make an overhead transparency (OHT) of Blackline Master 20 (or reuse the OHT from Activity 3).

Activity

1. The students should each have 10 triangles, a smaller pentagon, and a larger pentagon (see Activity 3). On the larger pentagon, have the students draw all the diagonal lines to join the vertices while you do the same on the OHT.

2. Label all of the points both on and inside the pentagon as shown in Figure A on the left, and have the students copy your actions.

3. With the students, identify a number of the individual triangles inside the pentagon. Describe each triangle by the points on each of its vertices. See *Triangle ABF* in Figure B, below left.

4. Discuss how larger versions of the two types of triangles can be seen if different shapes are combined. For example, Triangle ACD is a larger version of Triangle AGF. Say, *Find a small acute triangle in your pentagon. Measure the length of one of the longer sides. Now divide it by the length of the short side. Find a larger version of the acute triangle and repeat the steps. Compare the results. What do you notice?* The students should report that both answers are about 1.6.

5. Say, *Shapes that are the same shape but not the same size are said to be "similar". If shapes are the same, it means that corresponding angles in both shapes are equal. For these triangles, it also means that the ratio of the long side to the short side is the same, regardless of size.*

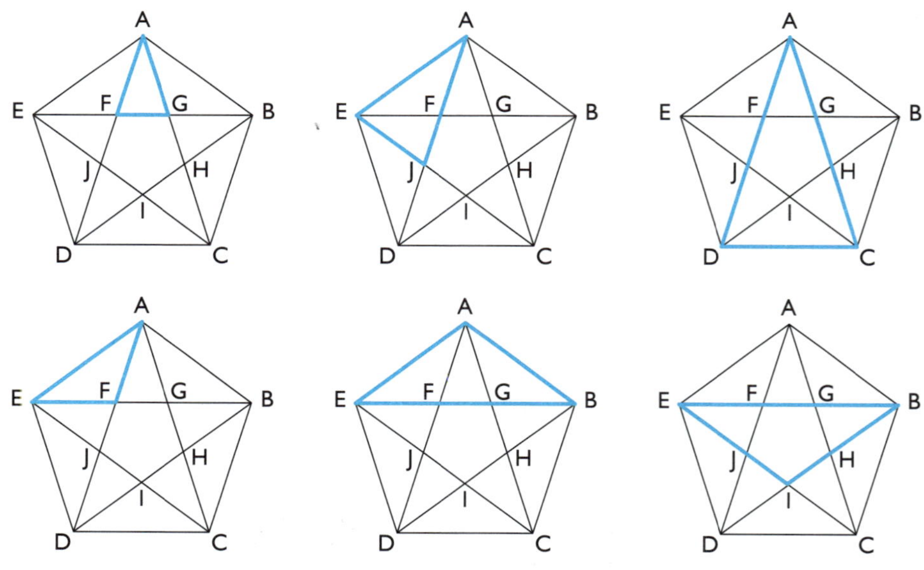

▲ *A number of mathematically similar shapes can be found inside the pentagon.*

6. Have the students also investigate the ratio of similar obtuse triangles that can be seen in the pentagon. They will discover that the ratio between the long and short sides is also approximately 1.6 : 1.

7. Instruct the students to compare the long side of the obtuse triangle with the long side of the acute triangle. They will find that, once again, the ratio between these sides is about 1.6 : 1, as is the ratio between the short sides of the obtuse and acute triangles.

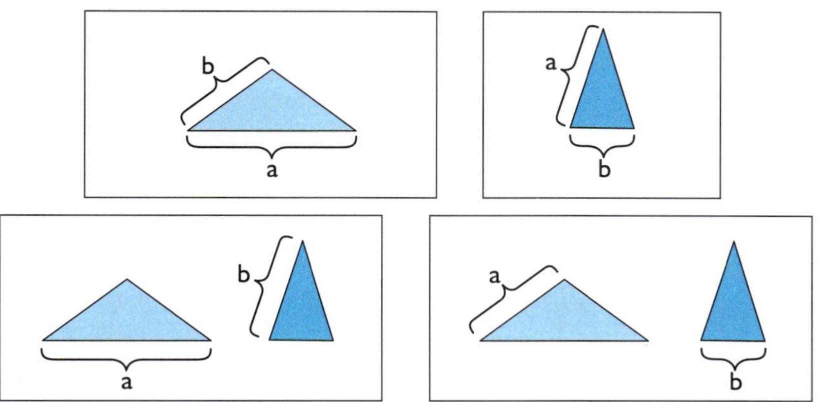

▲ *The ratio between (a) and (b) in all of these figures is about 1.6 : 1.*

Did You Know?

An acute triangle is one in which the largest angle is acute (less than 90°).

An obtuse triangle is one in which the largest angle is obtuse (more than 90° but less than 180°).

geo An easy way of making mathematically similar shapes is by folding metric paper. Instructions for 15 different polygons can be found in *Paper Polygons*.

10. The Golden Ratio

Preparation

1. Copy Blackline Master 20 (page 78) onto the tagboard or light card.

2. Inside one of the pentagons, draw diagonal lines to connect the vertices and cut out the resulting shapes (or reuse the pieces made in Activity 5).

Activity

1. Review Activity 9 ("Finding Similar Shapes"). Discuss the ratio of the long side to the short side of each isosceles triangle within the pentagon. Highlight that the ratio that was continuously found was of approximately 1.6 : 1. Explain that this ratio is called the "Golden Ratio" (see the note section in the margin on page 58). Point out that the Golden Ratio appears in many other situations, for example, in a particular type of oblong.

2. By arranging the triangles on the overhead transparency, guide the students in creating Figure A (shown below) on their sheets of scrap paper. The students can keep their pieces in position with Blu-Tack.

■ Materials

- Blackline Master 20 (page 78)
- Tagboard or light card
- Scissors — for demonstration
- Overhead projector and blank transparency sheet
- Overhead transparency pen
- Students' pentagon pieces from Activity 3
- Blu-Tack
- Scrap paper — 1 sheet for each student

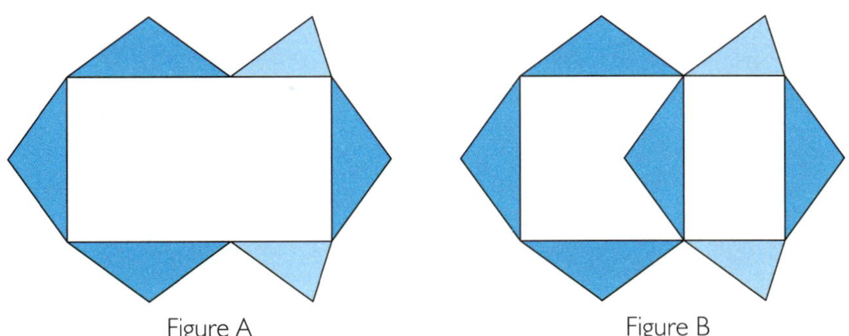

Figure A   Figure B

▲ *Arranging the triangles in this way can help students explore the Golden Ratio.*

An oblong is a rectangle with adjacent sides of different lengths. Also known as a "non-square rectangle".

A "vertex" is the point where two or more line segments join or intersect on a 2D shape.

Shapes that look exactly the same, apart from their size, are said to be "similar". Corresponding angles are also the same size.

The Golden Ratio has many different names, including "Golden Section" and "Divine Proportion", all of which reflect the high regard that the ratio has had throughout history. It is also known as phi (φ) or tau (τ). The oblong examined in Activity 10 ("The Golden Ratio") is called the "Golden Rectangle", and it can be seen in many examples of art and architecture. The acute triangle examined in Activity 9 ("Finding Similar Shapes") is known as the "Golden Triangle", while the obtuse triangle is known as the "Golden Gnomon". Examples of the Golden Ratio can also be found in nature. For example, the ratio between the distances from an adult's foot to the top of their head and from their foot to their navel is about 1.6 : 1.

3. Ask, *What is the ratio of the long side of this oblong is to the short side?* The students will probably guess the answer, but have them measure just to check. They will discover that the ratio is approximately 1.6 : 1.

4. Add another obtuse triangle, as shown in Figure B on page 57, and ask, *What two shapes can you see inside the oblong?* (A square and another oblong.) *Look at the smaller oblong. Measure the sides to figure out the ratio. What is it?* (About 1.6 : 1.)

5. Add the two acute angles and draw a line to connect the vertices, as shown below. Point out that there is now another square and oblong. Instruct the students to repeat the steps, and have them once again calculate the ratio of the sides of the new oblong (about 1.6 : 1). Say, *We could continue dividing each new oblong into smaller squares and oblongs forever, and the ratio of the long side to the short side of the new oblongs will always be approximately one and six-tenths to one.*

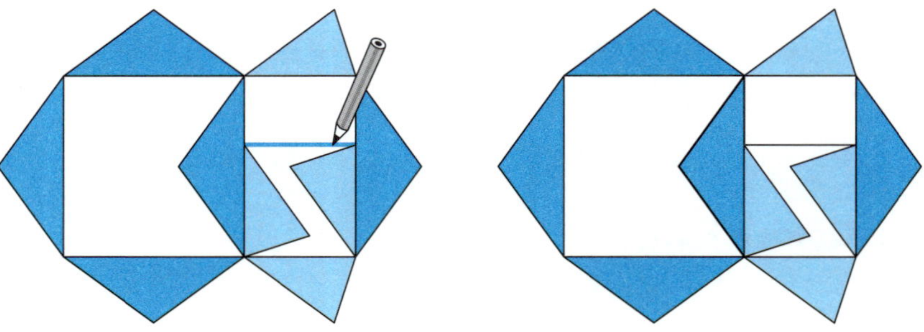

▲ *Draw a line between the vertices of the triangles to reveal another similar oblong.*

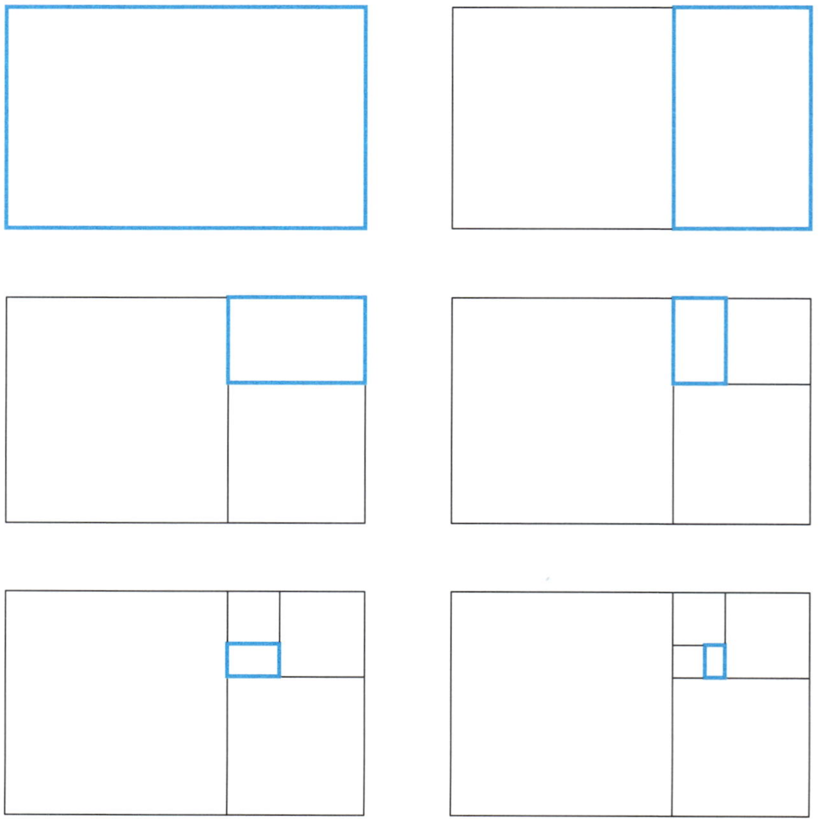

▲ *An oblong that uses the Golden Ratio can be continuously divided into smaller oblongs that have the same ratio.*

geo
shape cards

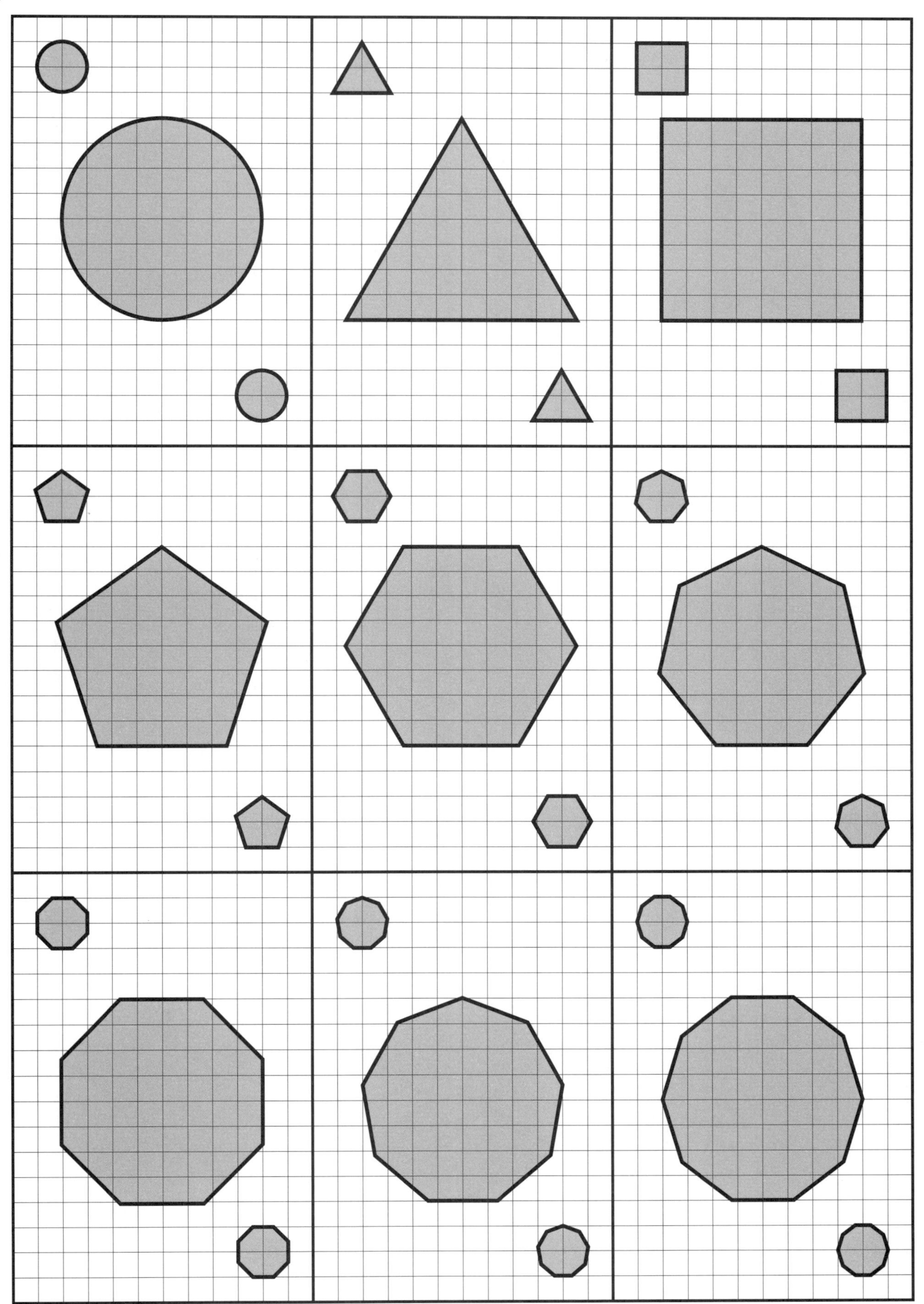

Blackline Master 2

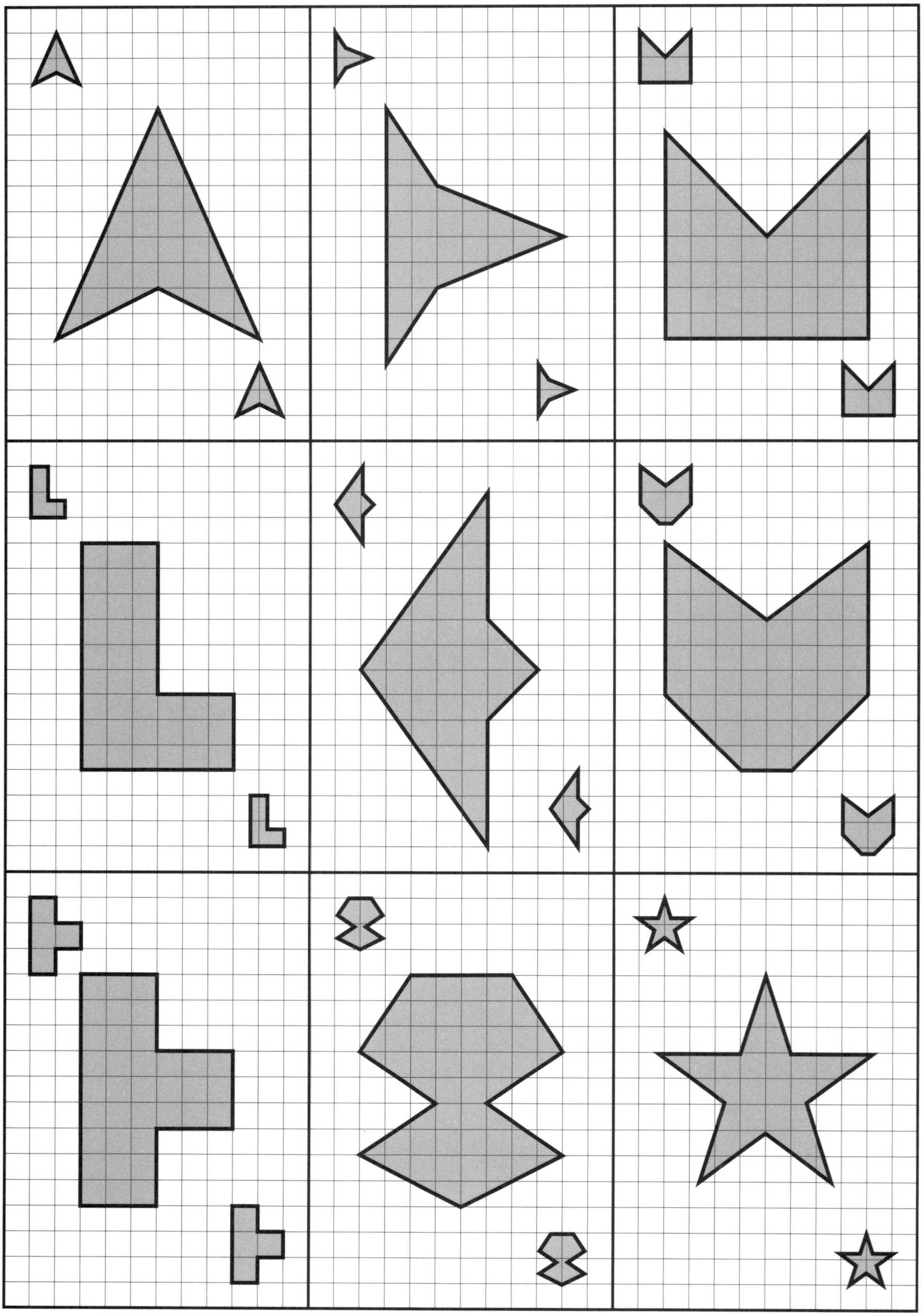

geo
shape cards

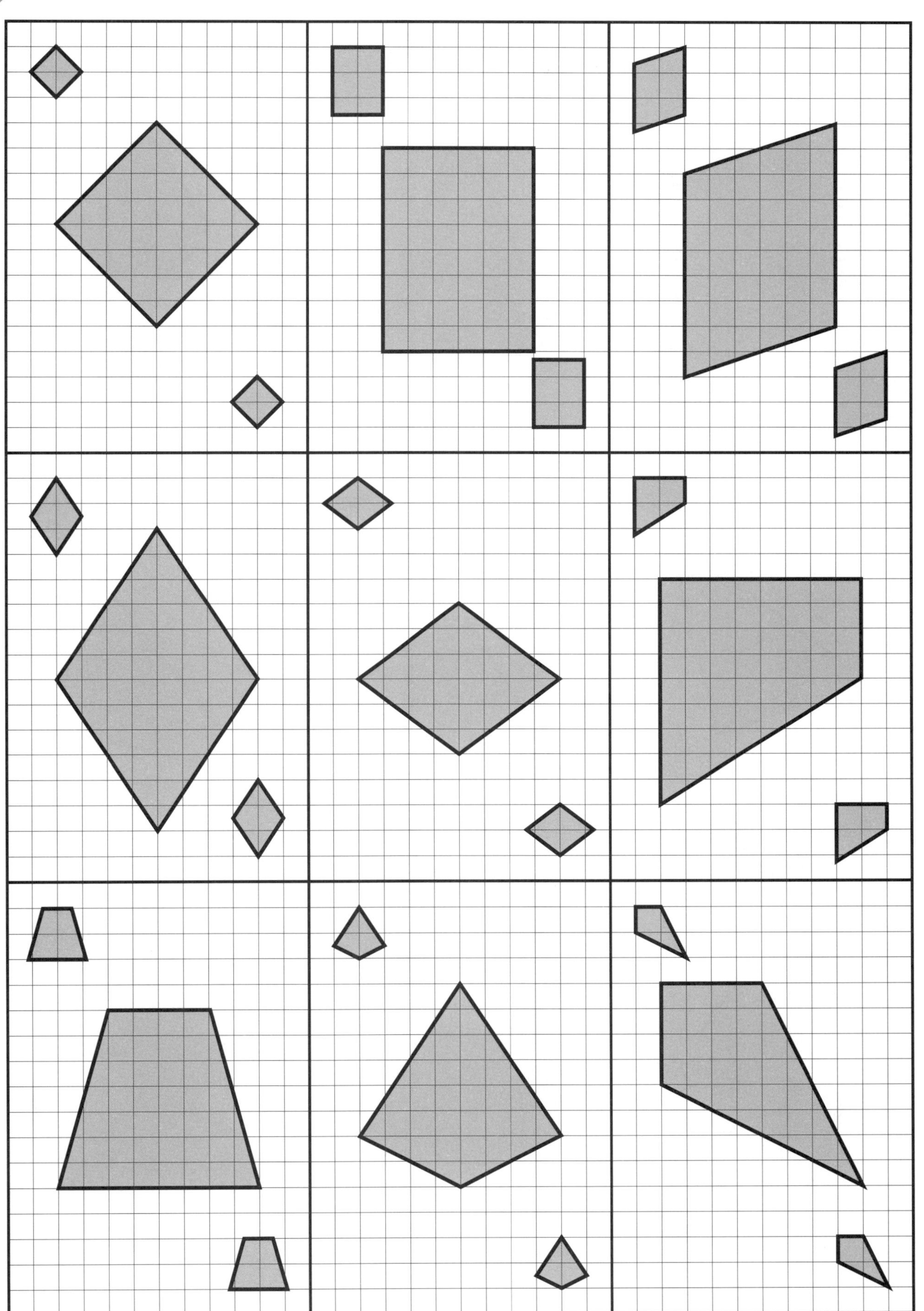

Blackline Master 3

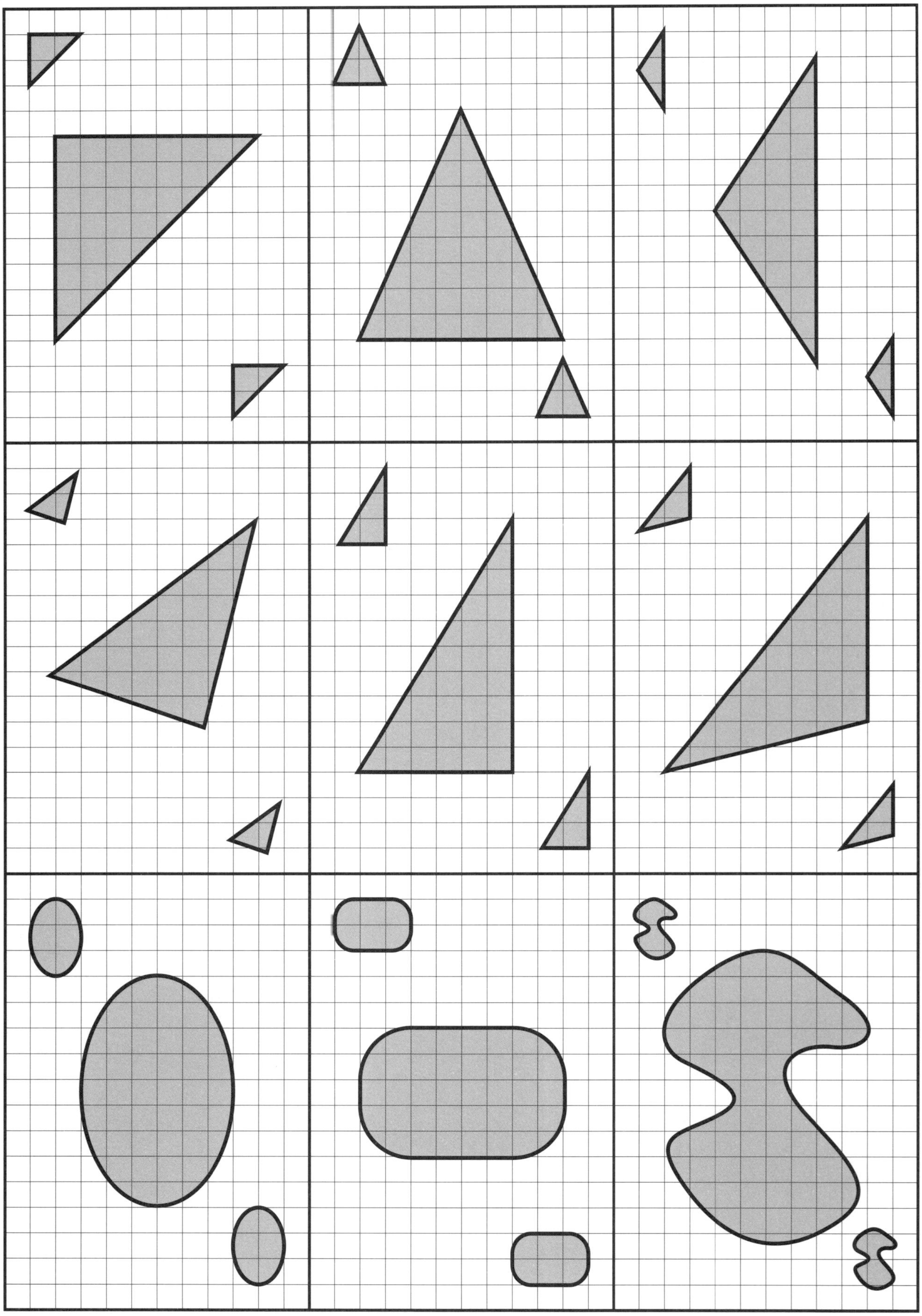

geo *shape cards*

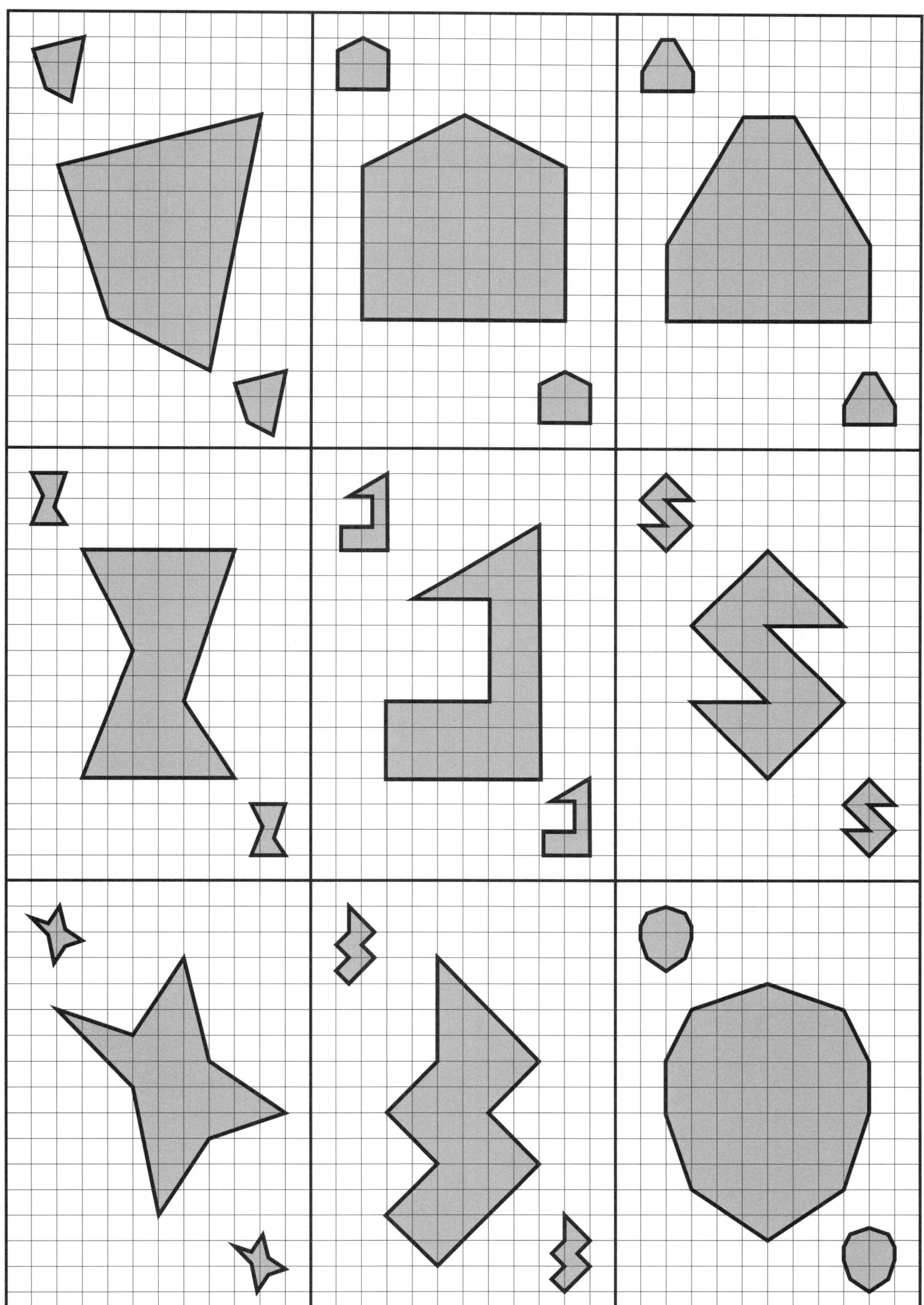

Blackline Master 5

why isn't a quadrilateral a quadrigon?

Why isn't a quadrilateral a quadrigon or a tetralateral, tetragon, quadrangle, or tetrangle?

Ever wonder why mathematics can seem like a foreign language? Many of the words that we use to describe 2D shapes come from ancient Latin or Greek words.

Polygons

The word "polygon" comes from Greek *polys,* meaning "many", and *gonia,* meaning "angle" or "corner". The origins of *gonia* may be related to an older Indo-European word *genu,* meaning "knee".

"Angle" comes from the Latin word *angulus,* meaning "a little bending". *Angulus* may have come from the Indo-European words *ang* or *ank,* meaning "to bend". A related English word is "ankle".

Quadrilaterals

- kite — named after the toy, which flies like a bird called a kite. Originally from Old English *cyta,* a type of hawk

- oblong — Latin *longus,* meaning "long"

- parallelogram — from three Greek words, *para,* meaning "alongside", *allenon* meaning "one another", and *gramma,* meaning "letter" or "figure"

- quadrilateral — Latin *quadri,* meaning "four", and *latera,* "side"

- rectangle — Latin *rect,* meaning "upright" or "perpendicular" and *angulus,* "a little bending"

- rhombus — Greek *rhombos,* an object used to make noise: it looked like the rhombus we use today in geometry

- square — Latin *exquadrare* meaning "to make square", from *quadra,* "four"

- trapezoid/trapezium — Greek *trapeza,* meaning "table"

Something-a-gons

Many mathematical terms use Greek and Latin prefixes. Polygons are named by combining the prefixes below with "gon". Look in a dictionary to see where else these prefixes are used.

1   *henos, mono* (Greek); *un, una, uni* (Latin)

2   *di, do, duo* (Greek); *duo* (Latin)

3   *tri* (Greek)

4   *tetra* (Greek); *quadri* (Latin)

5   *pent* (Greek)

6   *hex* (Greek)

7   *hept* (Greek); *sept* (Latin)

8   *oct* (Greek)

9   *ennea* (Greek); *nona* (Latin)

10   *deca* (Greek)

11   *hendeca* (Greek *henos* and *deca*)

12   *dodeca* (Greek *do* and *deca*)

Triangles

- acute — Latin *acutus,* meaning "pointed" or "sharp". Originally from *acus,* meaning "needle"

- equilateral — Latin *æquus,* meaning "equal", and *latus,* "side"

- isosceles — Greek *isos,* meaning "equal", and *skelos,* "leg"

- obtuse — Latin *obtusus,* meaning "blunted" or "dull"

- right — Old English *riht,* meaning "straight", later meaning upright or perpendicular

- scalene — Greek *skalenos,* "uneven"

geo compass creations

Equilateral Triangle

1. Draw line segment AB so that it is 3 cm long.

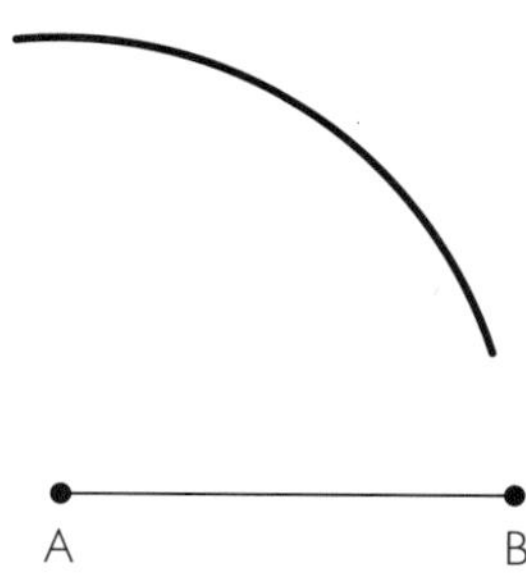

2. Set the compass to 3 cm. With the compass point on A, draw an arc.

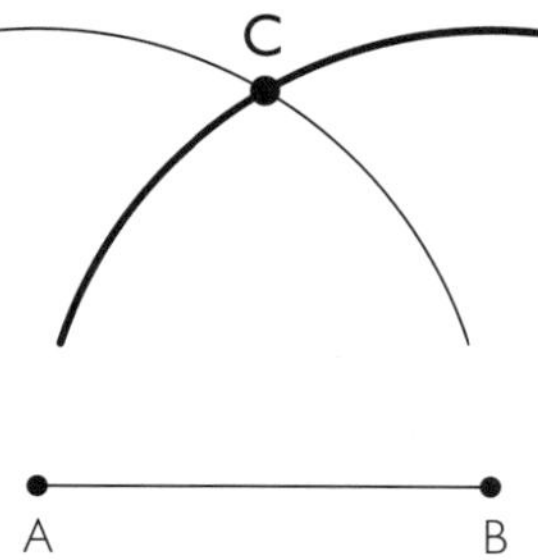

3. From point B, draw a second arc. Mark point C where the two arcs intersect.

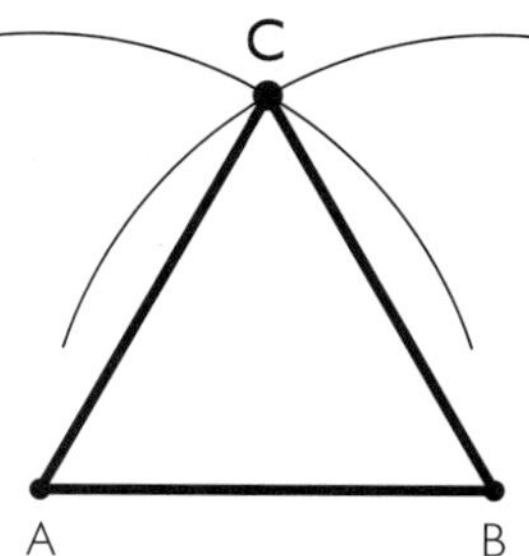

4. Join points A, B, and C.

Regular Hexagon

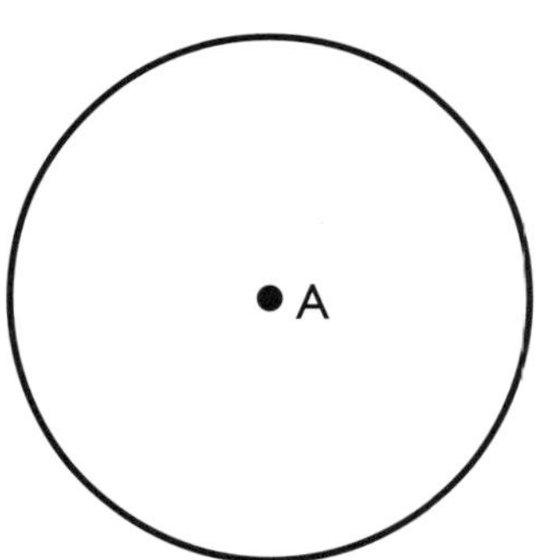

1. Set the compass to 2 cm for the rest of these instructions. Using point A as the central point, draw a circle.

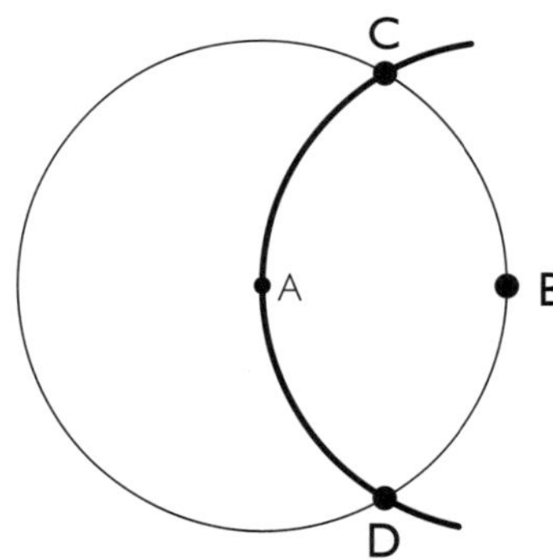

2. Mark point B anywhere on the circle. From point B, draw an arc. Mark points C and D where the arc and circle intersect.

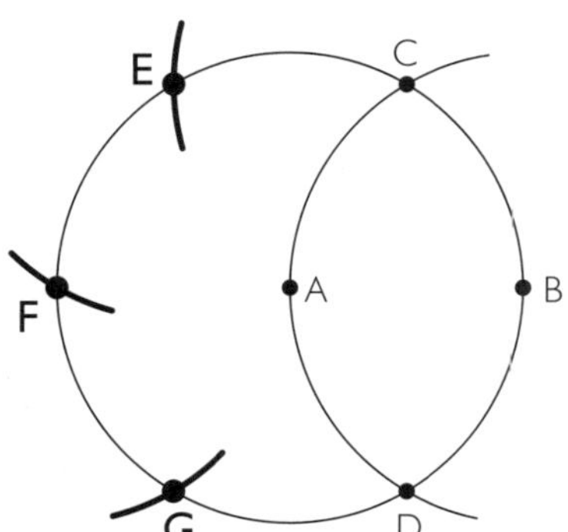

3. From point C, draw an arc to intersect the circle at point E. Repeat from point E to find F, and from F for G.

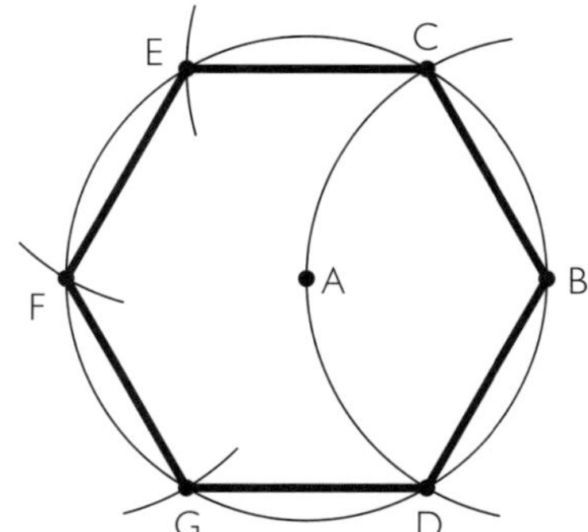

4. Join points B, C, E, F, G, and D.

geo compass creations

Kite

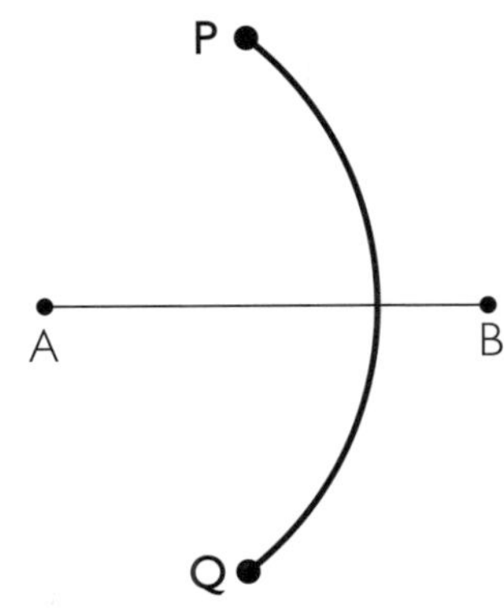

1. Draw line segment AB so that it is 4 cm long.

2. Set the compass to 3 cm. With the compass point on A, draw the arc PQ.

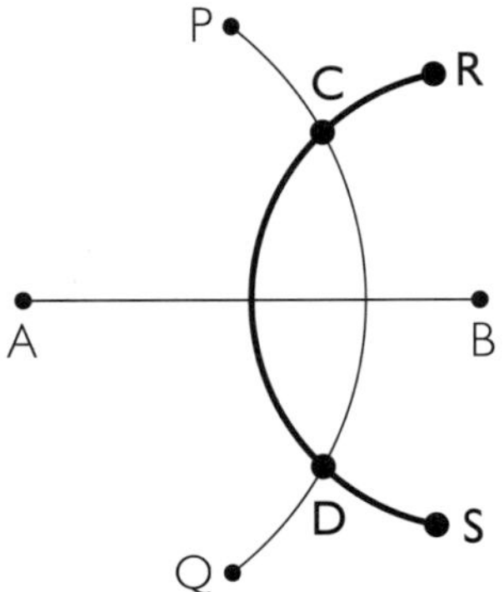

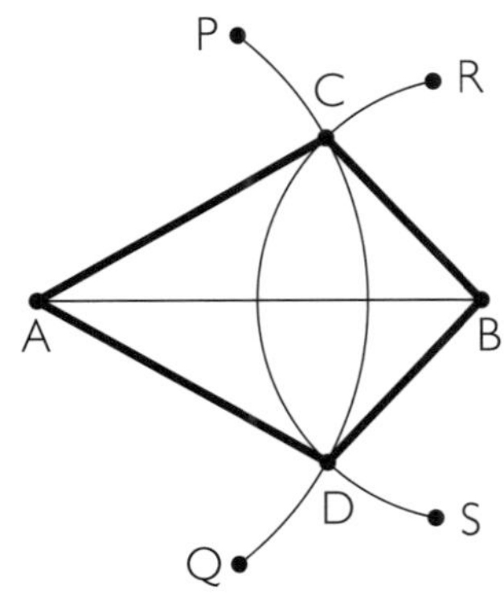

3. Set the compass to 2 cm. From point B, draw the arc RS. Mark points C and D where the two arcs intersect.

4. Join points A, C, B, and D.

Parallelogram

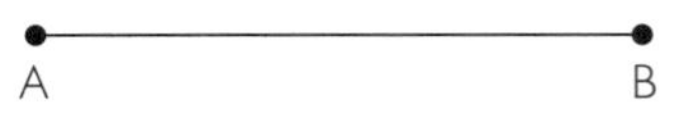

1. Draw line segment AB so that it is 4 cm long.

2. Set the compass to 3 cm. From point A, draw arc PQ. From point B, draw arc RS.

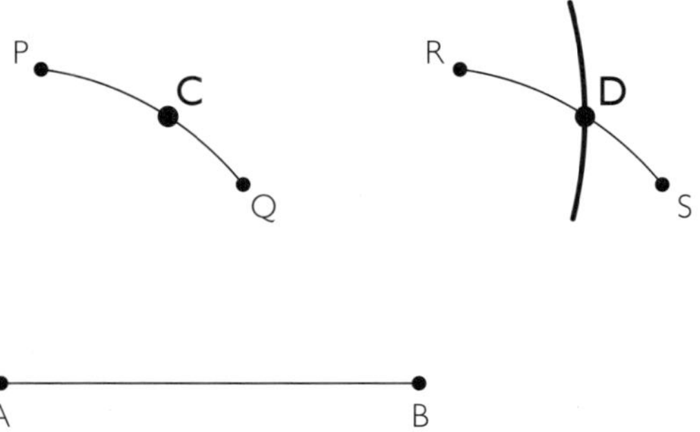

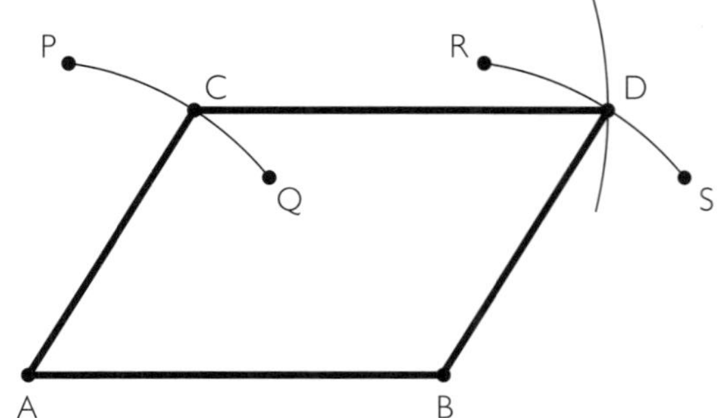

3. Mark point C anywhere on arc PQ. Set the compass to 4 cm. From point C, draw an arc to intersect arc RS at point D.

4. Join points A, C, D, and B.

Blackline Master 8

g̲e̲o̲ compass creations

Perpendicular Lines

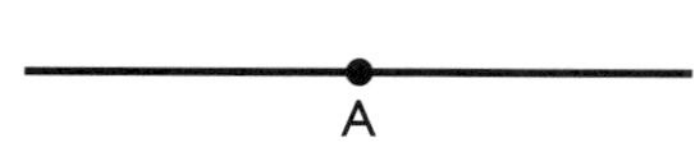

1. Draw a line segment and mark point A in about the middle.

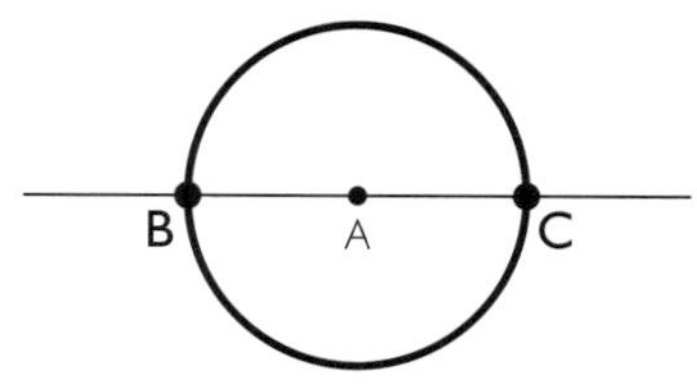

2. Using point A as the central point, draw a circle. Mark points B and C where the circle intersects the line segment.

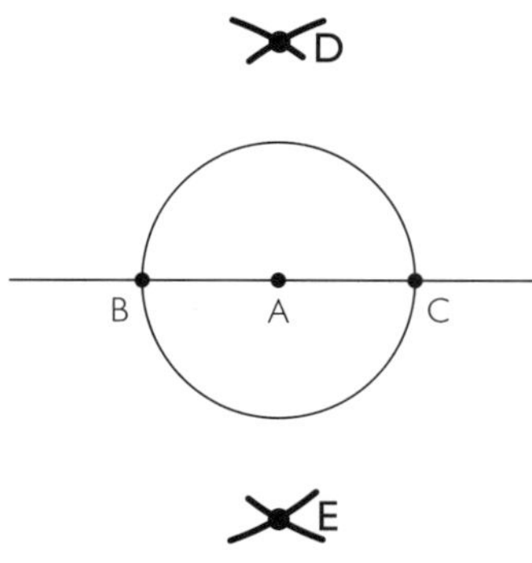

3. Set the compass to the distance between points B and C. From points B and C, mark arcs above and below the line segment to intersect at points D and E.

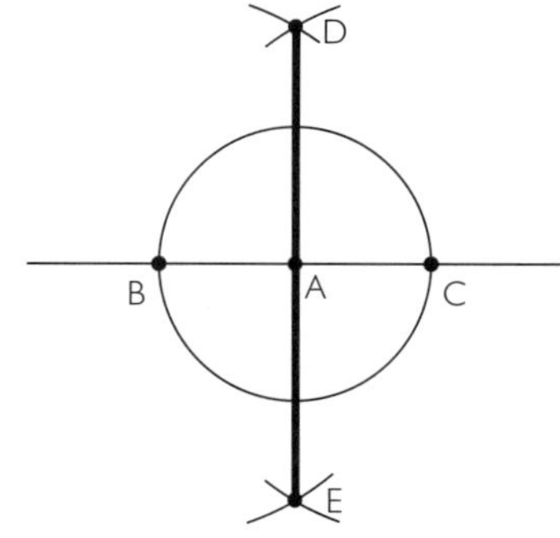

4. Join points D and E. The line segment DE is perpendicular to line segment BC.

Oblong

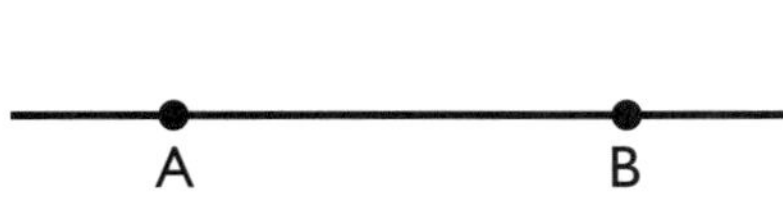

1. Draw a line segment and mark points A and B so that they are 4 cm apart.

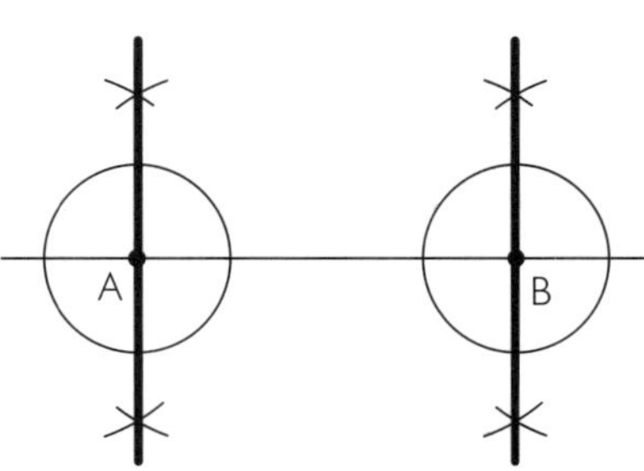

2. Construct perpendicular lines from points A and B.

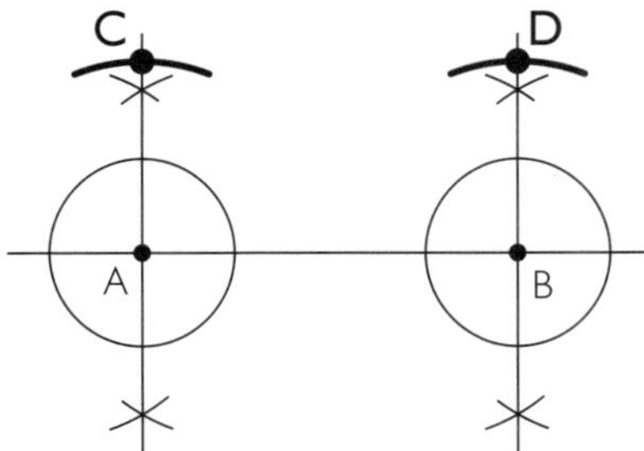

3. Set the compass to 2cm. From points A and B, mark arcs to intersect the perpendicular lines at points C and D.

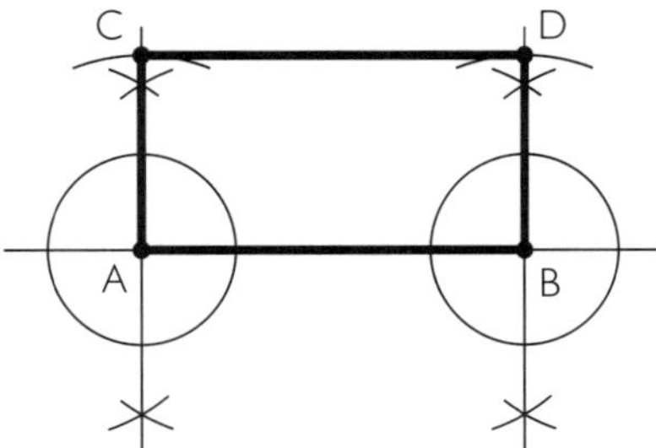

4. Join points A, C, D, and B.

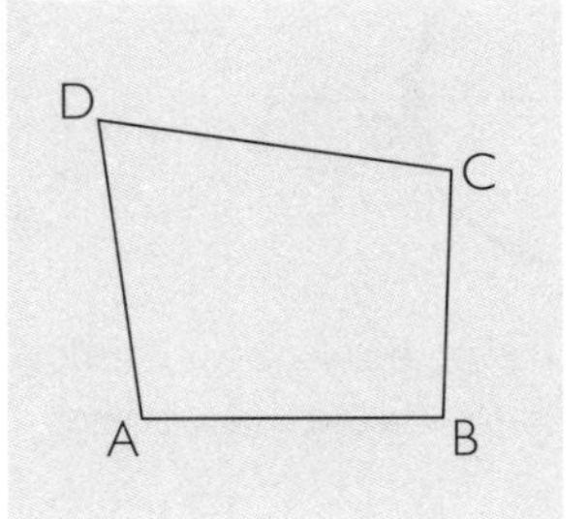
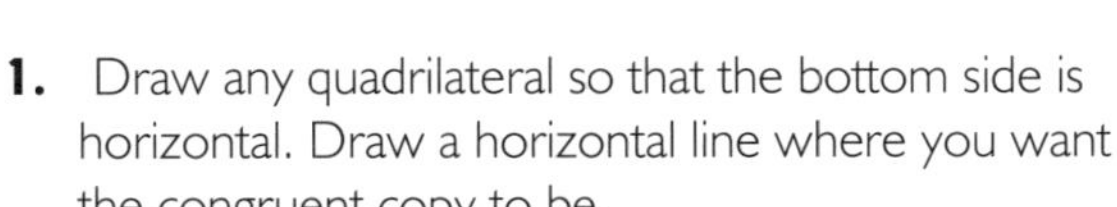
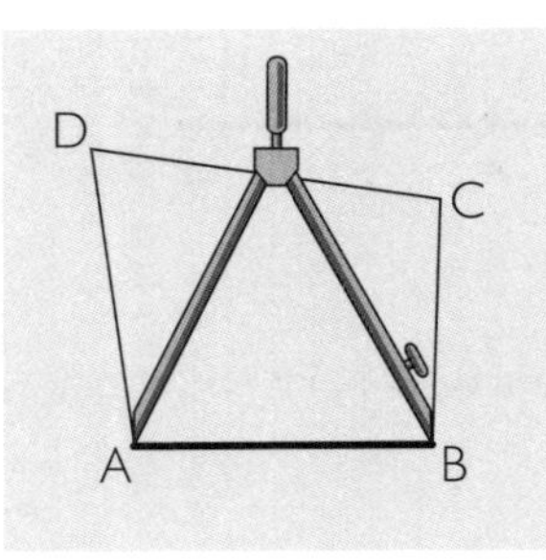
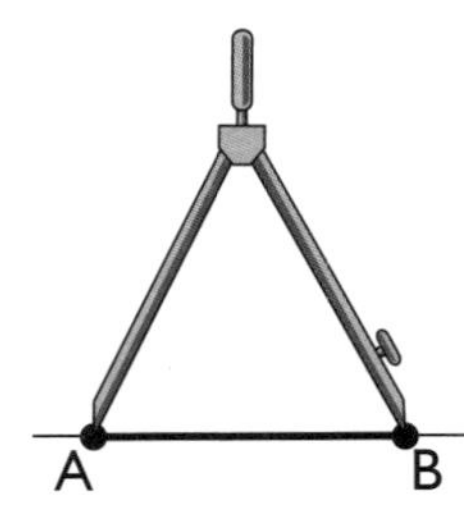

1. Draw any quadrilateral so that the bottom side is horizontal. Draw a horizontal line where you want the congruent copy to be.

2. Set the compass to the distance between points A and B. Mark this distance on the horizontal line and label the points.

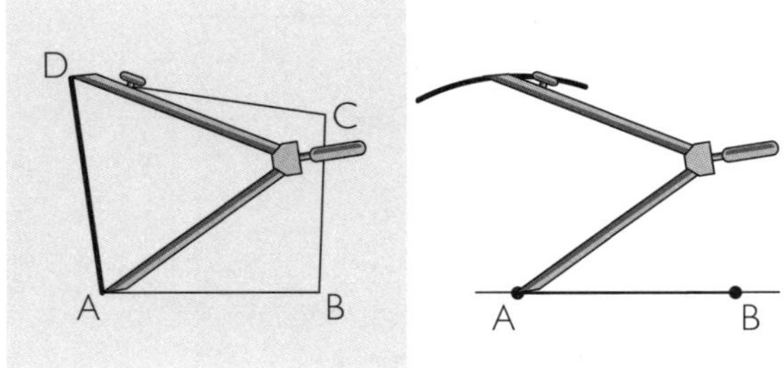
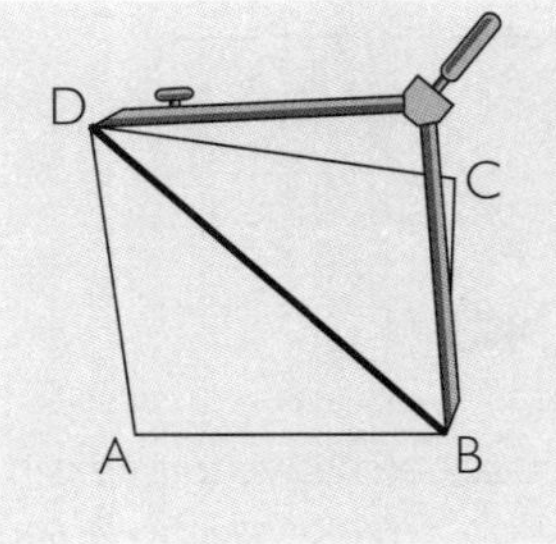
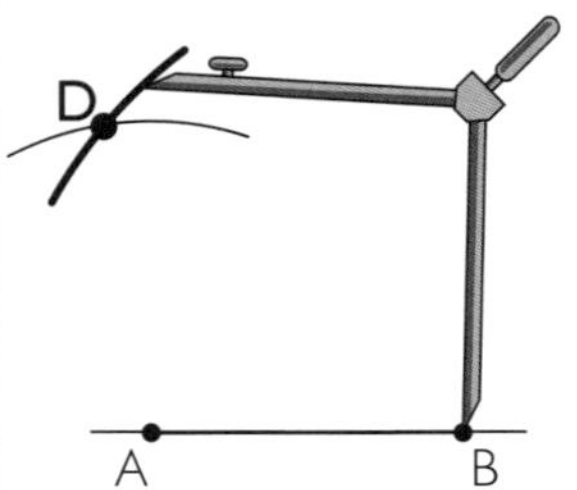

3. Set the compass to the distance between points A and D. Draw an arc to show this distance on the copy.

4. Set the compass to the distance between points B and D. Draw an arc to show this distance on the copy. Label the intersection point as D.

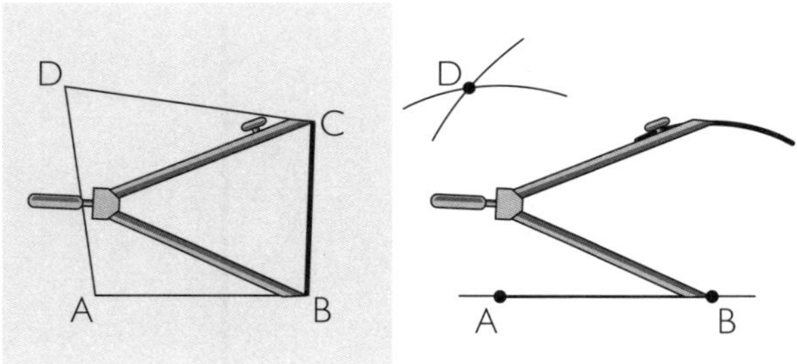
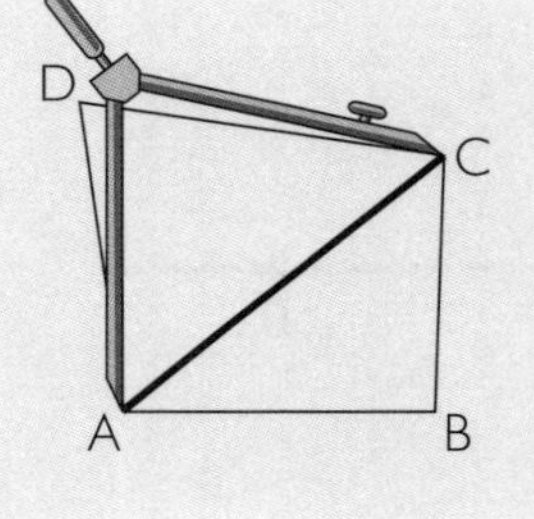
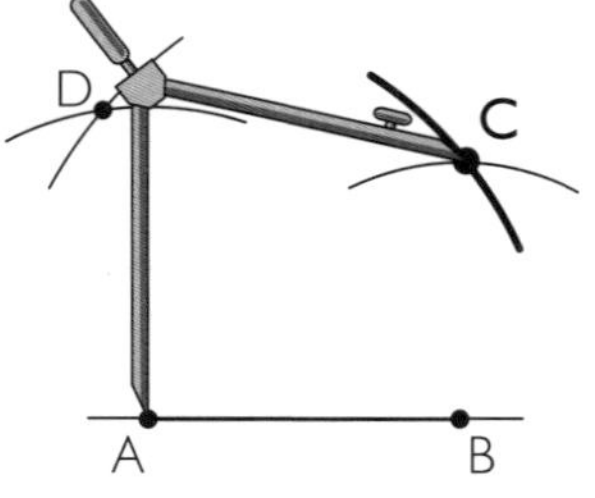

5. Set the compass to the distance between points B and C. Draw an arc to show this distance on the copy.

6. Set the compass to the distance between points A and C. Draw an arc to show this distance on the copy. Label the intersection point as C.

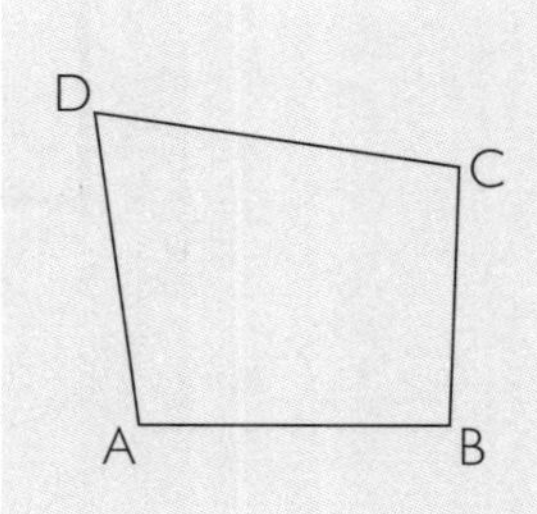
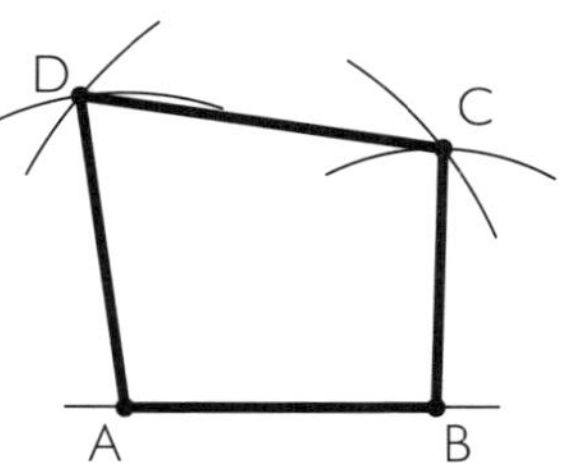

7. Connect all of the points on the copy to show the congruent shape.

1.

2.

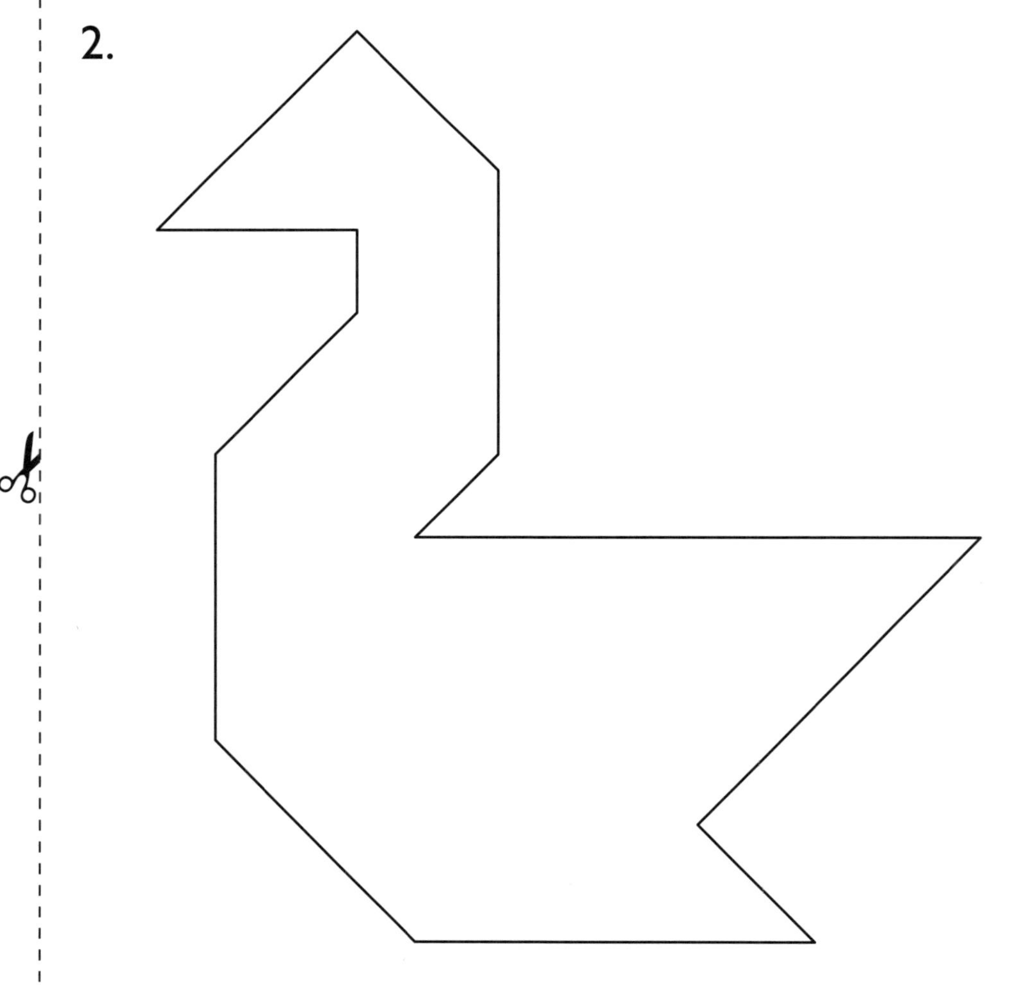

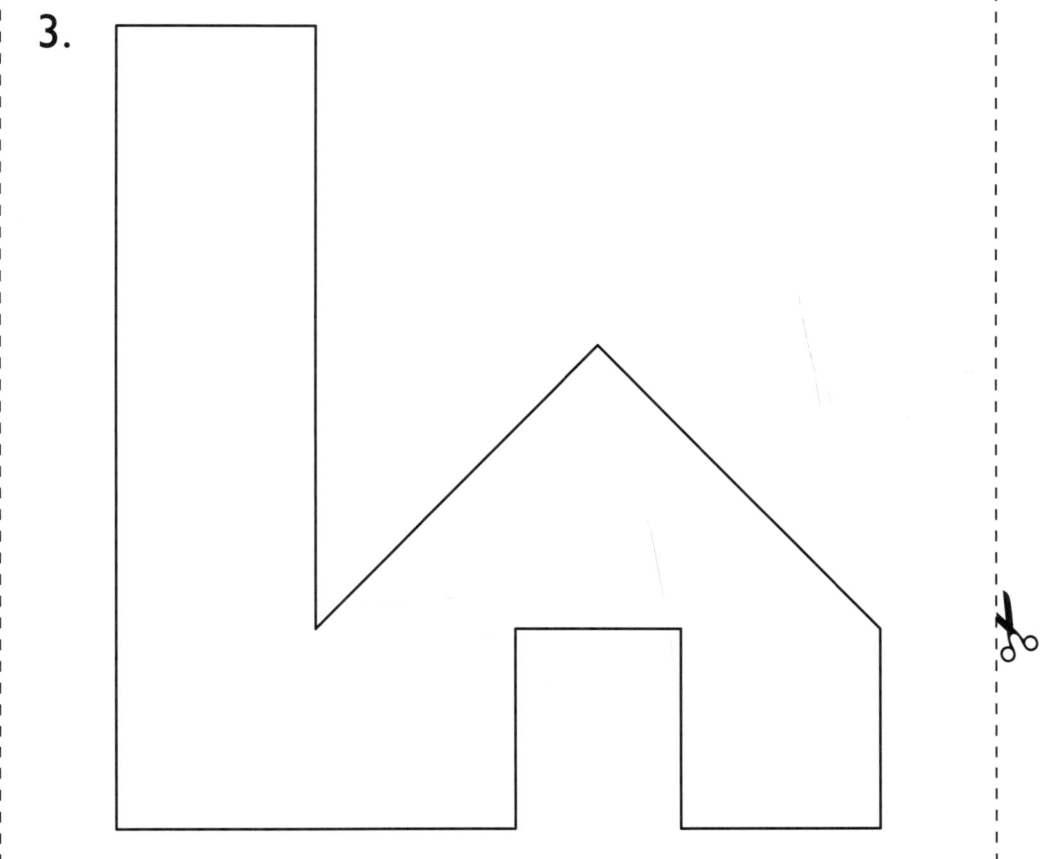

3.

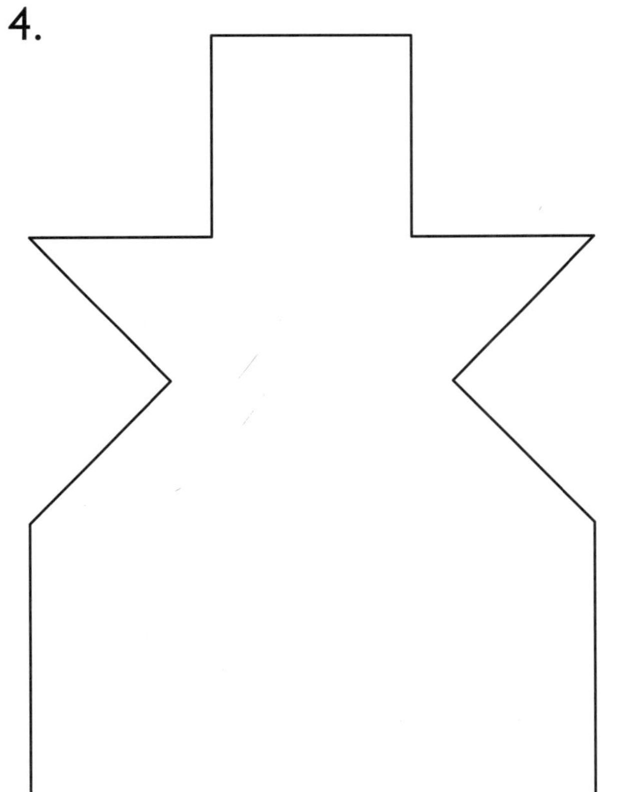

4.

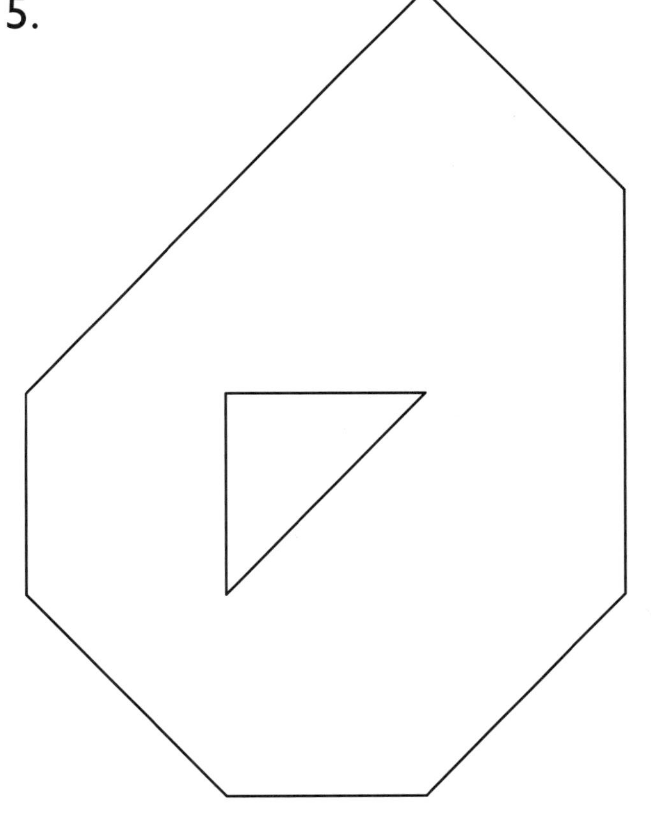

5.

# tangram puzzles

6.

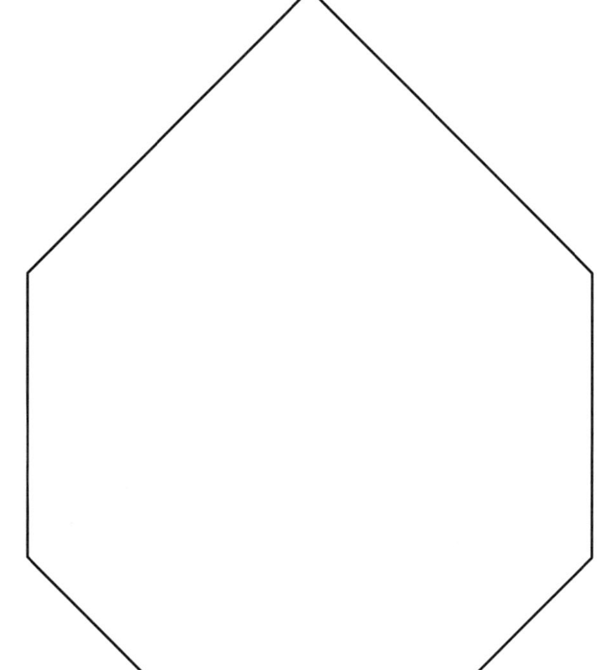

7.

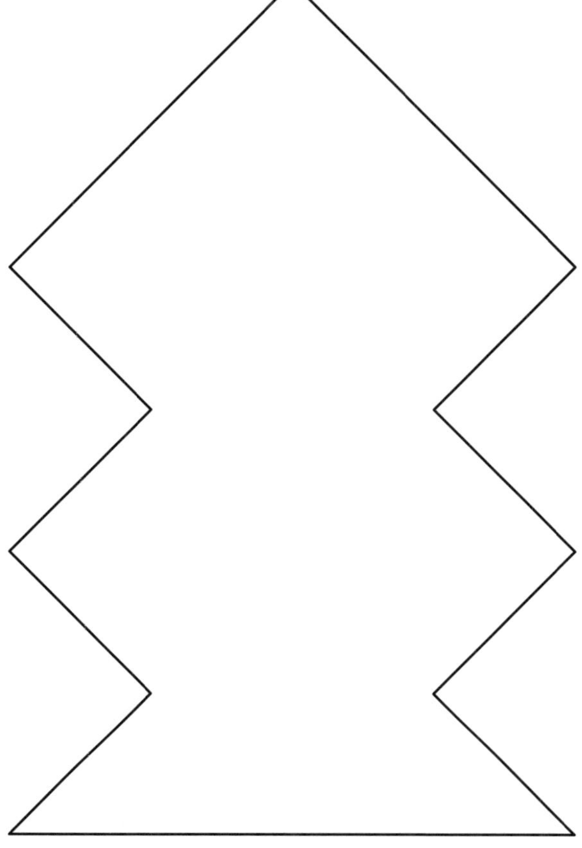

8.

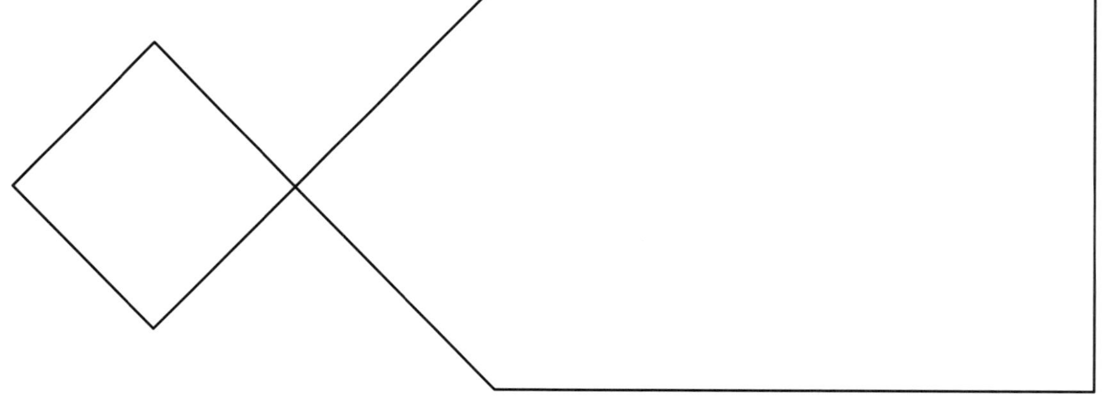

Blackline Master 14

geo polyhexes and polyquads

Dihex – 1

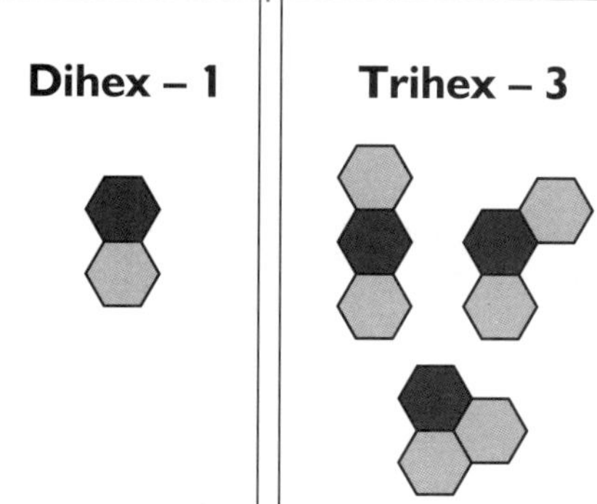

Trihex – 3

Tetrahex – 7

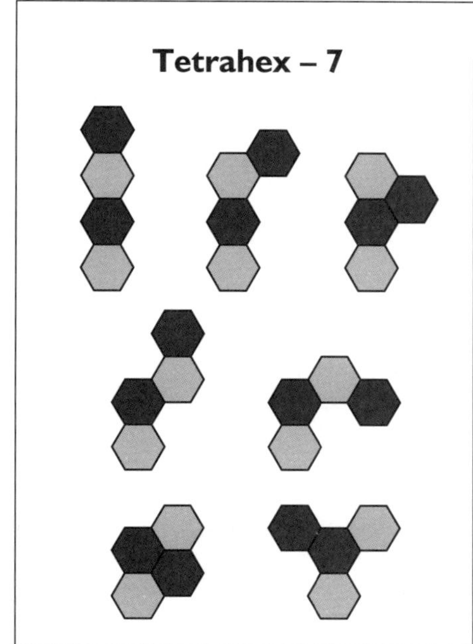

Pentahex – 22

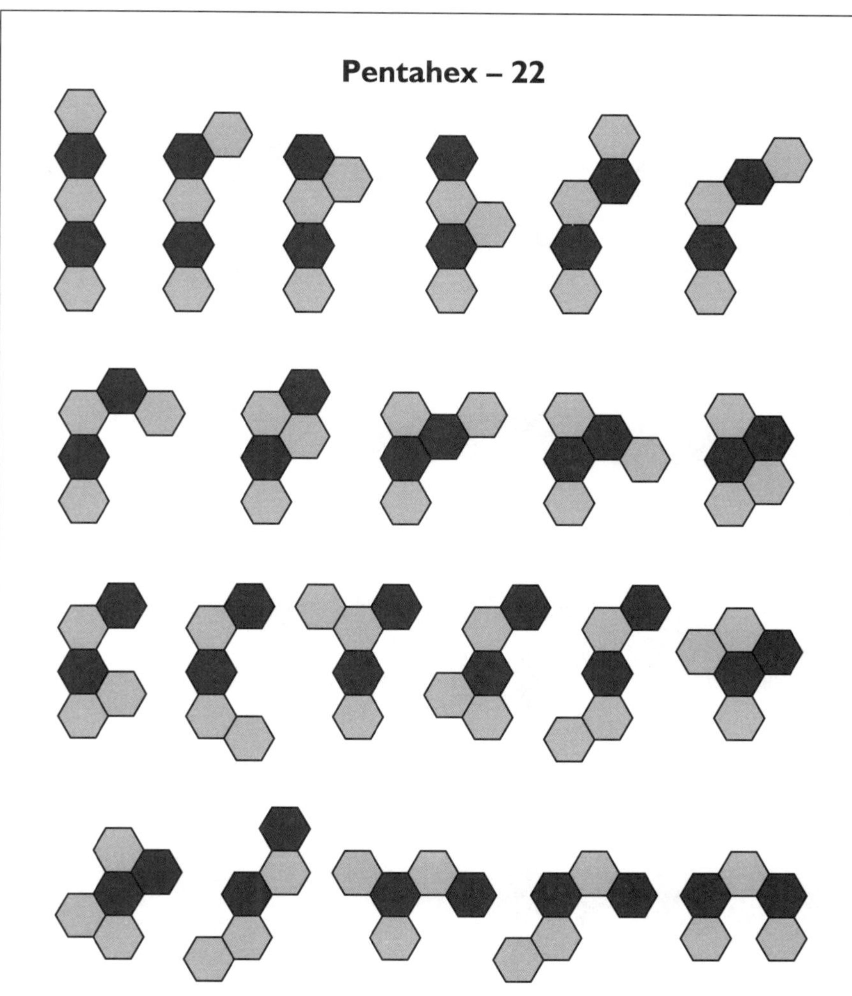

Diquad – 2

Triquad – 3

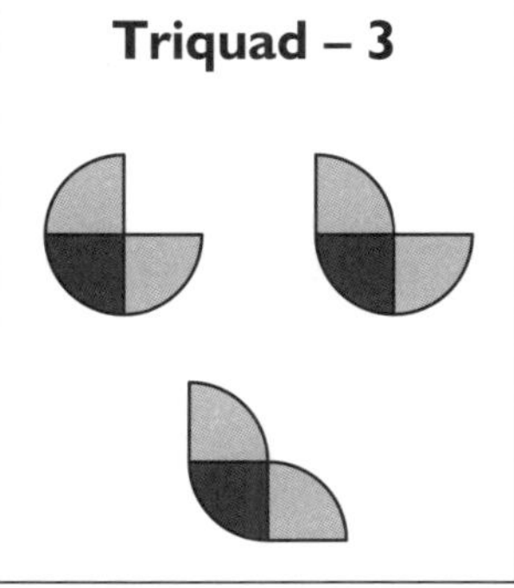

Pentaquad – 7

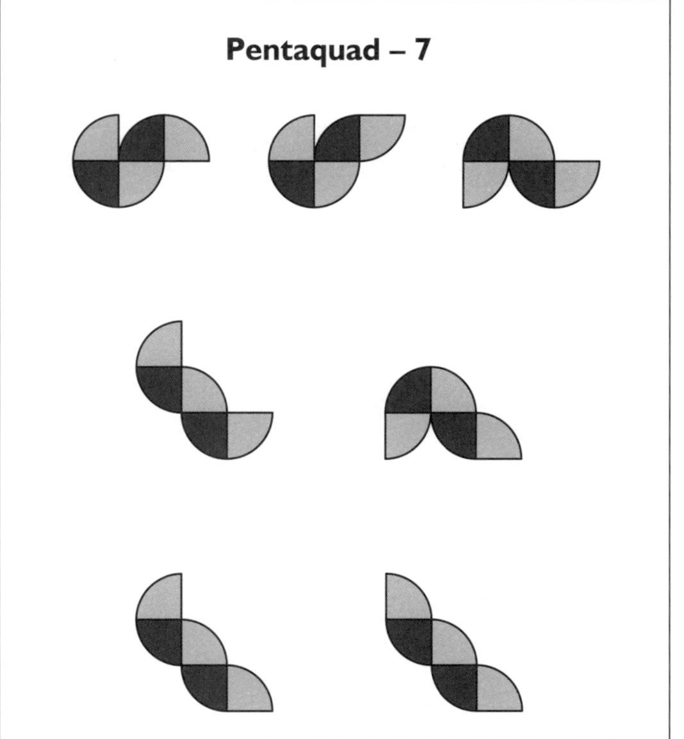

Tetraquad – 6

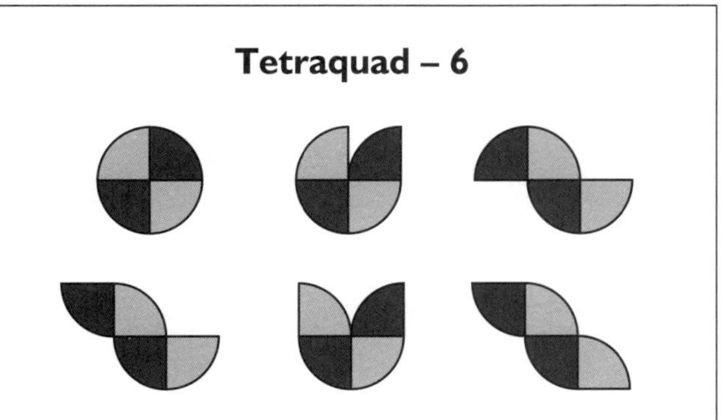

Blackline Master 19

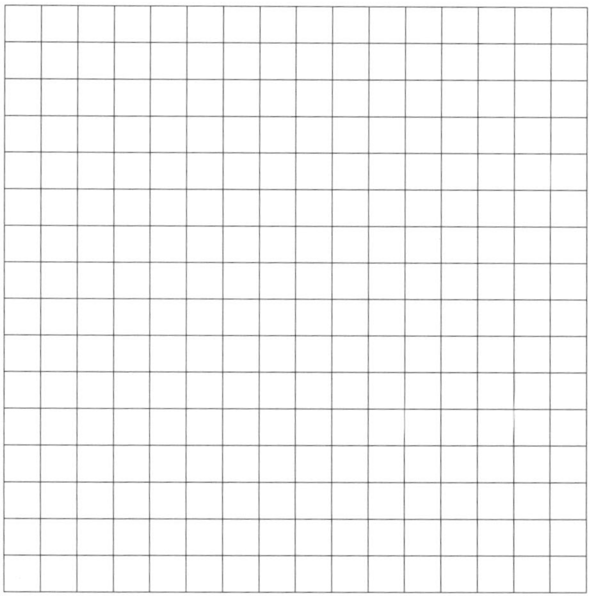

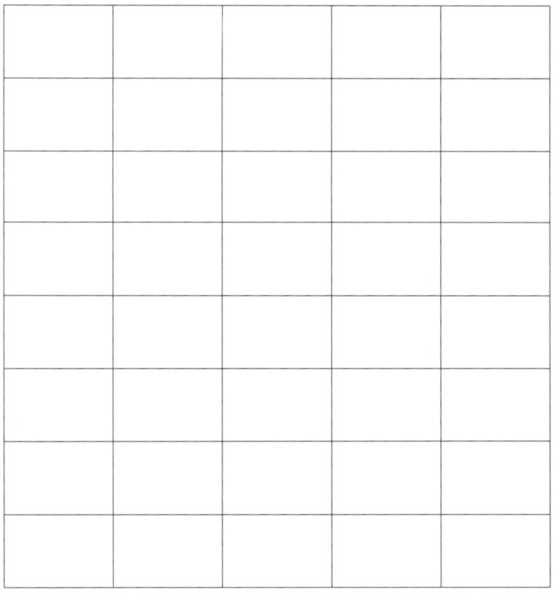

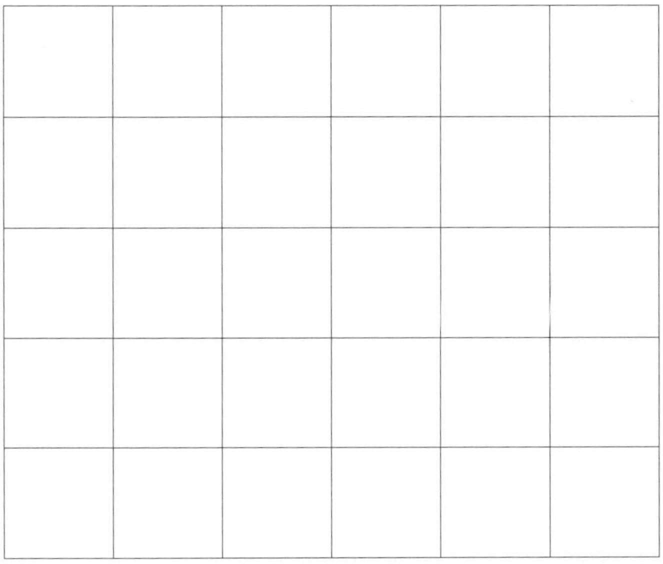

pentagons

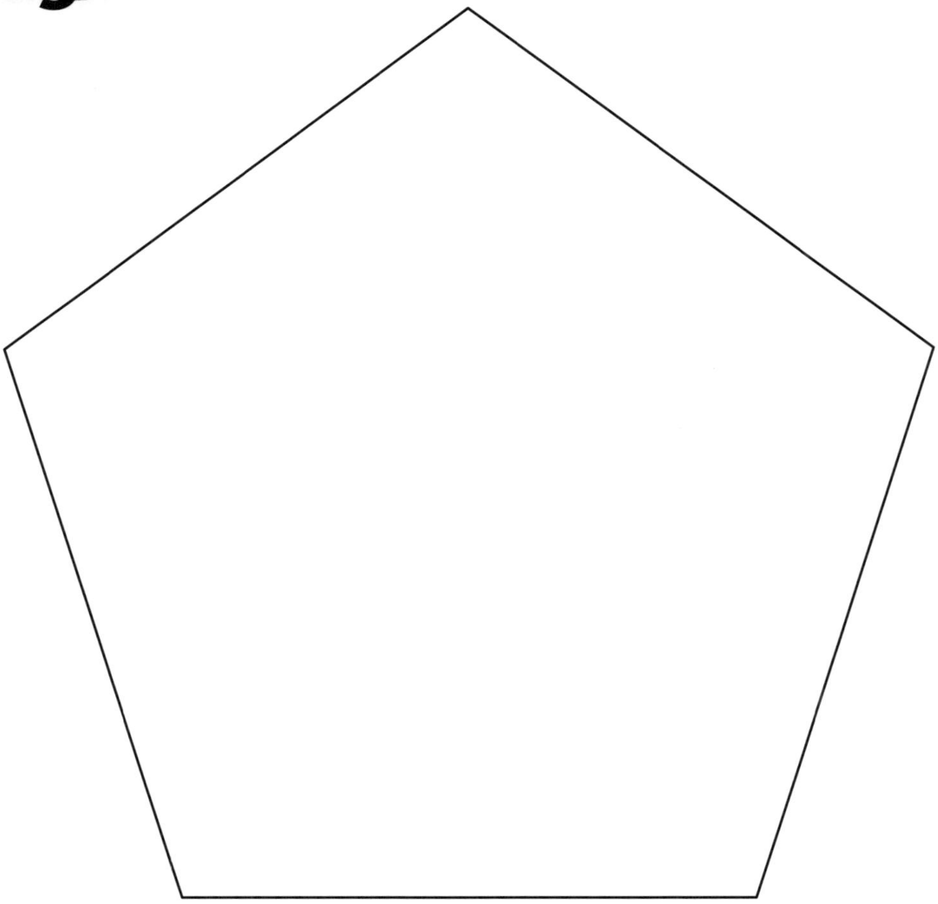

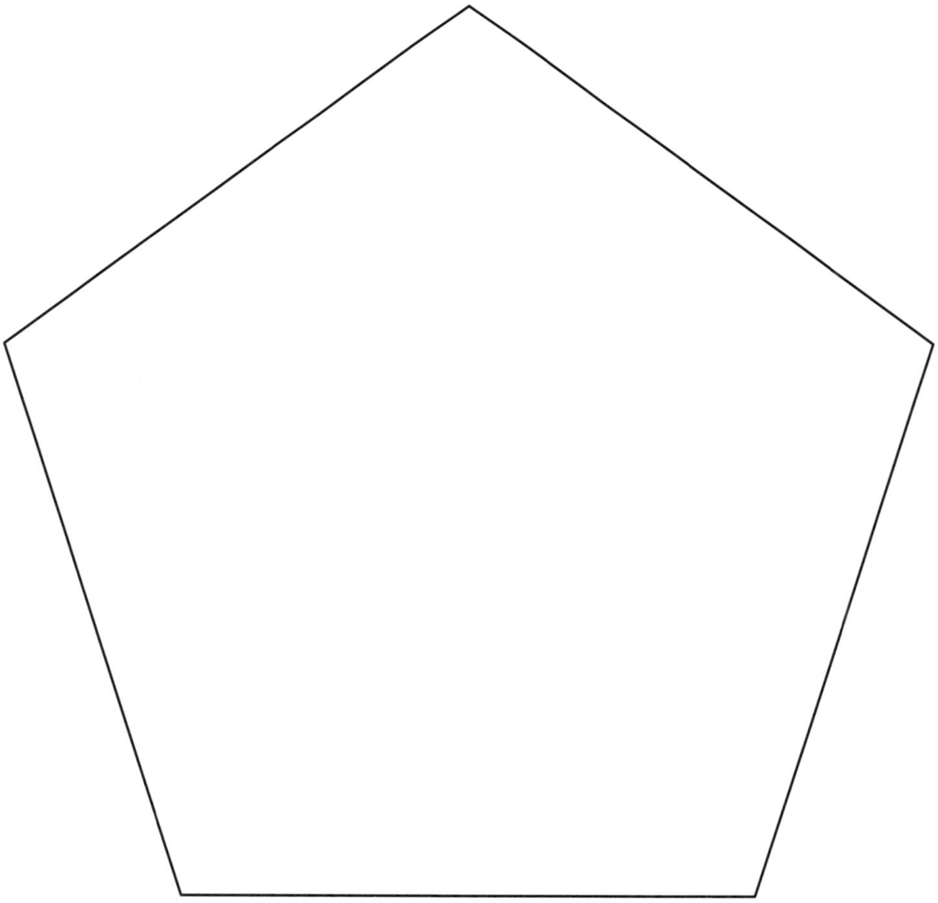

Blackline Master 20

glossary

Angle: When two line segments share the same vertex they create an angle. The size of the angle is determined by the amount of turn from one line segment to the other.

Arc: A part of any curve. Usually used to describe part of the circumference of a circle.

Circle: A closed two-dimensional curve formed by a set of points that are an equal distance from a central point. It can also be thought of as an ellipse that has its foci at a common point.

Closed two-dimensional shape: A shape that divides the plane into two regions: an interior and an exterior. A point in the interior joined to a point in the exterior will always intersect at least once with the shape.

Concave: Concave polygons have at least one interior angle that is larger than 180°.

Concentric: If two mathematically similar shapes have a common center they are said to be concentric.

Congruent: Shapes that are exactly the same size and shape are said to be congruent. Congruent angles are the same size.

Convex: The interior angles of a convex polygon are all less than 180°.

Decagon: Any polygon with ten sides. A regular decagon has ten equal sides and each angle is 144°.

Diagonal line: Any straight line segment joining two non-adjacent vertices in a polygon.

Diameter: Any line segment that passes through the center of a circle and joins two points on the circumference. A diameter is equal to two radii.

Dodecagon: Any polygon with twelve sides. A regular dodecagon has twelve equal sides and each angle is 150°.

Ellipse: A closed, two-dimensional curve formed by a set of points. The sum of the distances from any point on the perimeter to two special points (foci) inside the ellipse is always constant.

Equilateral triangle: A triangle with three equal sides. Each angle is 60°.

Hendecagon: Any polygon with eleven sides. A regular hendecagon has eleven equal sides and each angle is approximately 147.3°

Heptagon: Any polygon with seven sides. A regular heptagon has seven equal sides and each angle is approximately 128.6°.

Hexagon: Any polygon with six sides. A regular hexagon has six equal sides and each angle is 120°.

Horizontal: A line or surface that is completely level is called horizontal.

Irregular polygon: A polygon which does not have all sides equal, all angles equal, or both.

Isosceles trapezoid/trapezium: A trapezoid/trapezium in which the non-parallel sides are equal in length.

Isosceles triangle: A triangle with two sides of equal length and two equal angles. The angles opposite equal sides are equal in measure.

Kite: A quadrilateral with two pairs of equal adjacent sides, where one pair is different in length to the other pair. Only one pair of opposite angles is equal.

Line: A straight line extends in either direction for an infinite distance into space. In everyday use, it usually refers to a line segment.

Line segment: A line segment is part of a straight line and has two end points. In everyday use, it is usually called a line.

Nonagon: Any polygon with nine sides. A regular nonagon has nine equal sides and each angle is 140°.

Oblique: A line or surface which is neither vertical nor horizontal is called oblique.

Oblong: A rectangle with adjacent sides of different lengths. Also known as a "non-square rectangle".

Octagon: Any polygon with eight sides. A regular octagon has eight equal sides and each angle is 135°.

Open two-dimensional shape: An open shape is one where there is a gap in the boundary of the shape.

Parallel: When two lines are the same distance apart along their entire lengths, they are said to be parallel.

Parallelogram: A quadrilateral with opposite sides equal and parallel. Opposite angles are also equal.

Pentagon: Any polygon with five sides. A regular pentagon has five equal sides and each angle is 108°.

Perpendicular: When two lines intersect to form a right angle they are said to be perpendicular to each other.

Plane: A flat surface extending in all directions infinitely.

Point: A position in space that is usually marked by a small dot.

Polygon: Any simple two-dimensional closed shape formed by three or more straight line segments.

Quadrant: A special sector of a circle. It is formed by an arc and two radii that form a right angle. It covers one quarter of the area bounded by a circle.

Quadrilateral: Any polygon with four sides.

Radius: A line segment that joins the center of a circle with any point on its circumference. The plural of radius maybe either radii or radiuses.

Rectangle: Any parallelogram with all angles equal to 90°

Regular polygon: A polygon with all angles and all sides equal. A regular polyhedron is a three-dimensional shape that features the same regular polygon on all its faces.

Rhomboid: A parallelogram with adjacent sides of different lengths. All angles are greater than or less than 90°.

Rhombus: A parallelogram with four equal sides.

Right angle: An angle that is equal to 90°.

Scalene triangle: A triangle with sides that are unequal in length.

Sector: A sector is the region bounded by two radii and an arc of a circle.

Semicircle: A special arc of a circle, whose endpoints are the endpoints of a diameter. It is also thought of as the arc and the diameter together.

Similar: Shapes that look exactly the same, apart from their size, are said to be similar. Corresponding angles are also the same size.

Simple: A simple two-dimensional shape is one in which no part of a shape's boundary intersects itself.

Square: A rectangle that has all sides equal in length.

Symmetry: A shape is said to have line or reflective symmetry if half of the shape is a mirror image of its other half. If a shape matches its original outline after being rotated through a fraction of a full turn, then it is said to have rotational symmetry.

Tessellation: Shapes that are used to cover a plane to make a repeating pattern without any gaps are said to tessellate. The result is called a tessellation.

Trapezoid/trapezium: A quadrilateral with exactly one pair of parallel sides that are of different lengths. In North America, it is known as a trapezoid. Many other countries call the shape a trapezium.

Triangle: Any polygon with three sides.

Vertex: The point where two or more line segments join or intersect on a two-dimensional shape, or where three or more edges meet on a 3D shape.

Vertical: A line or surface that forms a 90° angle with a horizontal line or surface is called vertical.